Cambridge Elements ☰

Elements in the Foundations of Contemporary Physics
edited by
Richard Dawid
Stockholm University
James Wells
University of Michigan, Ann Arbor

QUANTUM GRAVITY IN A LABORATORY?

Nick Huggett
University of Illinois Chicago

Niels Linnemann
University of Geneva

Mike D. Schneider
University of Missouri

CAMBRIDGE
UNIVERSITY PRESS

CAMBRIDGE
UNIVERSITY PRESS

Shaftesbury Road, Cambridge CB2 8EA, United Kingdom

One Liberty Plaza, 20th Floor, New York, NY 10006, USA

477 Williamstown Road, Port Melbourne, VIC 3207, Australia

314–321, 3rd Floor, Plot 3, Splendor Forum, Jasola District Centre,
New Delhi – 110025, India

103 Penang Road, #05–06/07, Visioncrest Commercial, Singapore 238467

Cambridge University Press is part of Cambridge University Press & Assessment,
a department of the University of Cambridge.

We share the University's mission to contribute to society through the pursuit of
education, learning and research at the highest international levels of excellence.

www.cambridge.org
Information on this title: www.cambridge.org/9781009327534
DOI: 10.1017/9781009327541

First published 2023

A catalogue record for this publication is available from the British Library.

ISBN 978-1-009-32753-4 Paperback
ISSN 2752-3039 (online)
ISSN 2752-3020 (print)

Quantum Gravity in a Laboratory?

Elements in the Foundations of Contemporary Physics

DOI: 10.1017/9781009327541
First published online: June 2023

Nick Huggett
University of Illinois Chicago

Niels Linnemann
University of Geneva

Mike D. Schneider
University of Missouri

Author for correspondence: Nick Huggett, huggett@uic.edu

Abstract: The characteristic energy scale of quantum gravity makes experimental access to the relevant physics apparently impossible. Nevertheless, low-energy experiments linking gravity and the quantum have been undertaken: the Page and Geilker quantum Cavendish experiment, and the Colella–Overhauser–Werner neutron interferometry experiment, for instance. However, neither probes states in which gravity remains in a coherent quantum superposition, unlike – it is claimed – recent proposals. In essence, if two initially unentangled subsystems interacting solely via gravity become entangled, then theorems of quantum mechanics show that gravity cannot be a classical subsystem. There are formidable challenges to such an experiment, but remarkably, tabletop technology into the gravity of very small bodies has advanced to the point that such an experiment might be feasible in the near future. This Element explains the proposal and what it aims to show, highlighting the important ways in which its interpretation is theory-laden.

Keywords: gravitationally induced/mediated entanglement, quantum gravity phenomenology/experiment, philosophy of experimental physics, quantum information theory and gravity, semiclassical gravity

ISBNs: 9781009327534 (PB), 9781009327541 (OC)
ISSNs: 2752-3039 (online), 2752-3020 (print)

Contents

1 Introduction

Physicists have searched for a fundamental theory of quantum gravity (QG) for nearly a century. Despite much progress on the theoretical side – including whole avenues of research toward how to develop such a theory (e.g. loop quantum gravity or string theory), little has been claimed on the empirical side. According to standard lore, this is entirely unsurprising given the Planck energy scale compared to the energy scales probed in high-energy particle physics; Zimmermann (2018) for instance pointedly illustrates the remoteness of the Planck energy scale for our usual collider technologies based on acceleration of charged particles in electric and magnetic fields:

> An ultimate limit on electromagnetic acceleration may be set by the Sauter-Schwinger critical field, above which the QED vacuum breaks down. ...Assuming these fields, the Planck scale of 10^{28} eV can be reached by a circular or linear collider with a size of about 10^{10} m, or about a tenth of the distance between earth and sun, for either type of collider (!). (pp. 36–37, exclamation mark in original)

An astrophysical benchmark may further help to communicate just how remote the physics in question is from our more familiar empirical world: The phenomenon of Hawking radiation, by whose detection we would like to corroborate the formal apparatus of quantum field theory in curved spacetime (merely *on the way to* a theory of QG apt for the Planck energy scale) is so weak that "Trying to detect astrophysical Hawking radiation in the night's sky is thus like trying to see the heat from an ice cube against the background of an exploding nuclear bomb" (Thébault 2016, p. 4).

What is the empirically minded QG researcher to do? Despair not being an option, maybe desperation is: One could search for an evidential or confirmatory framework that leaves room for *non*empirical forms of support for developments on the theoretical side of QG research. From this perspective, the framework for nonempirical theory confirmation developed by Dawid (2013) in the context of string theory may be attractive. Alternatively, one could follow the subcommunity of *quantum gravity phenomenologists*: Those empirically minded QG researchers who have not stopped working on finding empirical signatures of QG, despite awareness of the naive estimate of the difficulties as that of, say, Zimmermann already quoted here. This is the tack we intend to take.

One strategy in quantum gravity phenomenology is to look for QG effects within traces of high-energy astro-particle phenomenology in the early universe (famously motivated by using the universe as the "poor man's accelerator"). Another strategy is to systematically search for effects that "cascade" from high energies to low energies, such as in many cases of Lorentz invariance violation (in either the astrophysical-cosmological arena or in more controlled experimental settings). On this strategy, one accepts that the relevant energy scale is the Planck scale, but rejects a tacitly assumed fact of decoupling (Amelino-Camelia [2013] for a review).[1] A third strategy though, which will become our focus here, has only recently become relevant. It begins by noting that the Planck mass, rather than the Planck energy, might better serve as the quantity of interest in probing the quantum nature of gravity. As Christodoulou and Rovelli (2019) write:

> Puzzling is the fact that – unlike Planck length and Planck energy – m_{Planck} falls within a very reachable physical domain: micrograms. It has long been hard to see what sort of quantum gravity effect can happen at the scale of the weight of a human hair. (p. 65)

It has long been hard, but perhaps it will not be so hard any longer, and QG effects might indeed be in reach of tabletop experiments. (And if not literally tabletop, at least not solar-system sized!) Or at least, this is what recent claims amount to, in the emerging experimental research program known as *tabletop quantum gravity*.

It is important to note that thus restricting attention to the weak-field, Newtonian regime involves a significant change in the object of empirical study: One is no longer probing fundamental QG, only the low energy physics that different fundamental theories likely have in common. The question that faces us then is how might we read distinctively quantum traces of gravitational physics in such experiments? Answering this question seems to be key in making sense of the nascent tabletop quantum gravity research program. And in fact, only recently has it become clear that there is significant dissent among those physicists interested in quantum gravity phenomenology over the answer to this question.

---

[1] What of the "decoupling theorem" of quantum field theory? This strategy in quantum gravity phenomenology explores QG effects beyond field theory in the relevant sense.

A new call to interpret hypothesized results in a proposed class of tabletop quantum gravity experiments, which Bose et al. (2017) and Marletto and Vedral (2017b) have each independently noted may soon be viable, brings the question to the fore.

In this new class of *gravitational induced entanglement* (GIE) experiments (sometimes, Bose-Marleto-Vedral experiments; sometimes, gravitationally mediated entanglement experiments – names to be explained in Section 4.2) one employs spatially separated pairs of "gravcats,"[2] or gravitationally coupled Schrödinger cats (macroscopic, uncharged massive bodies) placed in spatial superposition, as the relevant quantum matter probes. Within these experiments, the hypothesized role for the underlying *quantum nature* of Newtonian gravity is to mediate entanglement between the two gravcats in a pair. The proposal, then, is that if such experiments indeed produce the predicted gravcat entanglement, then this outcome would provide a first ever laboratory *witness* of the quantum nature of gravity. And while this achievement would not amount to a direct observation of QG (and especially not to a direct observation that would distinguish between various current approaches to developing a fundamental theory of QG), it would still be an enormous advance. Yet, even the nature of this achievement in terms of a first tabletop quantum gravity witness is questioned by some in the community.

The stage thus set, three philosophers of physics came together, hoping to clear up for themselves a puzzle. How could it be that this specific, newly proposed class of experiments in tabletop quantum gravity could be a locus of dispute when all those involved in the dispute would seem to agree on their expected outcomes? And what do those results have to do with the supposed underlying *quantum* nature of gravity, anyway? This Element is our best attempt to provide a satisfying, unified answer to both of these interrelated questions. It is an answer that the three of us are, finally (after considerable friendly disagreement), content each to call our own.

The result (instead of a series of idiosyncratic articles written by each of us in turn, arguing back and forth through a thicket of distractions) is the following discursive work. It intertwines a review of the relevant

---

[2] The name "gravcats" goes back to Anastopoulos and Hu (2015).

physics suitable for philosophers of physics and physicists looking for a sketch of the field, with philosophical analysis giving both philosophers and physicists (whether internal or external to the field) a framework for understanding the conceptual and epistemic issues. We have thus aimed to provide the formal and philosophical background necessary to make everything comprehensible to our intended audience(s).

Following this Introduction, we offer a theoretical prelude (Section 2) on *semiclassical gravity* – a bit of theoretical architecture relevant for research in quantum gravity phenomenology, but whose conceptual status within our current best physics is equivocal. We distinguish three views of semiclassical gravity, including two different ways to deny that semiclassical gravity is itself to be understood as a candidate for future fundamental theory in the discipline, given today's best physics. In parallel with these two denials, we then provide in Section 3 an experimental prelude, noting two experiments in the early history of tabletop quantum gravity that are by now unavoidable in any conversation about quantum matter probes in a Newtonian gravitational context. Crucially, we will explain how these two experiments are importantly distinct from each other – particularly in the kinds of conclusions drawn from their successful execution. This observation occasions our identifying two traditions of experimental testing that will become relevant in our assessment of the GIE experiments, beginning in the section thereafter: On one tradition, the goal is to witness the quantum nature of gravity; on the other, one rather is interested in control of (or access to) the same.

With these preludes in place, we turn then to the GIE experiments, and our analysis spans the remaining four sections before the conclusion. After a preliminary naive rehearsal of the GIE experiments in Section 4, including a discussion of how they indeed would rule out semiclassical gravity as a candidate fundamental theory, we then turn to comment on the central question at stake when the experiments are considered in the tradition of witnessing (Sections 5–6): To what extent may we take the experiments as capable of witnessing a quantum nature of the gravitational state? We ultimately argue that one's answer to this question very much hinges on a choice of modeling "paradigm" (a term we use carefully) for the GIE experiments, even given agreement about fundamental physics. In particular, we develop what we have come to understand are the two major such paradigms in play in the relevant literature: what we call the "Newtonian model" paradigm and "tripartite models" paradigm, respectively. Only according to the latter does gravcat entanglement actually witness the quantum nature of gravity. Finally, in Section 7, we shift gears to offer the suggestion that the GIE experiments may at least as well be conceived in terms of their standing in the tradition of controlling and accessing – rather than witnessing – the quantum nature of gravity. And then we conclude.

Taking a step back, our first goal is thus to inspect the claim that a GIE experiment could amount to a tabletop quantum gravity witness, in light of the dissensus found within the physics community. We will find, on disentangling the various threads in the literature, that there is meaningful ambiguity as to whether the predicted outcomes of these experiments, if successful, would indeed provide such a witness. In particular, one's assessment depends on how one chooses to model the experimental setup, while our current best physics provides justifications for two distinct choices. However, our second goal is to provide a view of the GIE experiments that we believe adequately captures their payoff, as a matter at the frontiers of experimentation in tabletop quantum gravity, and which critically does not depend on choice of paradigm. Our hope in writing this Element is thus also to clarify that the successful completion of a tabletop quantum gravity experiment would be an enormous achievement for the experimental research program, regardless of further disagreements regarding the matter of witnessing.

2 Theoretical Prelude: "Semiclassical Gravity"

The problem of QG is generally understood as a need to unify two elements of our current corpus of fundamental physics: on the one hand, a classical and geometrized theory of gravity, general relativity (GR), which recovers Newtonian gravitation in an appropriate limit, while on the other hand, a quantum theory of (special) relativistic matter, the standard model of particle physics, which recovers nonrelativistic quantum mechanics (NRQM) in a different limit. But matter explicitly appears in GR as classical, contrary to our simultaneous embrace of a quantum field theory (QFT) description of matter, in the form of the standard model of particle physics – hence, the problem.

Of course, when seeking to unify a conflicted corpus, physicists typically proceed by trying to hold onto what are believed to be its crucial insights. In the case of QG, one obvious reconciliatory strategy begins with the observation that perhaps it is no requirement of GR that matter have a *fundamentally* classical nature. Rather, perhaps, at least for the sake of phenomenological modeling, there exists an *effective* classical description of the quantum matter – a classical stress-energy tensor. In the context of QFT, it is natural to associate any such effective classical quantities with expectation values of quantum observables, where (in a Hilbert space representation of the quantum state space) the latter are modeled as operators. Thus, one arrives at the Møller–Rosenfeld equation, dating back to the early 1960s (Møller et al. 1962; Rosenfeld 1963):

$$G_{\mu\nu} = \frac{8\pi G}{c^4} \langle \hat{T}_{\mu\nu} \rangle. \tag{2.1}$$

That is, the Einstein tensor $G_{\mu\nu}$, familiar from GR, couples to the expectation value of (what is now) a *quantum* stress-energy tensor operator, understood to act on any given prepared state of matter. (2.1) thus modifies the Einstein field equation of classical GR, replacing the stress-energy tensor on the right-hand side with its expectation value derived in quantum theory.

It is worth stressing that, despite looking (perhaps) innocuous as a modification to the Einstein field equation from GR, the Møller–Rosenfeld equation is a substantive conceptual departure from the classical equation. In the first place, whereas the left-hand side features a quantity that is meant to be descriptive of a single system, the right-hand side appears to describe a statistical property of a whole ensemble of systems. To see the point, imagine a version of (2.1) in which the right-hand side is an expectation value of a classical quantity, denoted by the same brackets, but no hat: It would describe a system in which each run of an experiment had the same Einstein tensor, determined by the statistical average of the different stress-energy tensors found in each run. In other words, the geometry on any particular run would depend on the stress-energy of all past and future runs, though only those that somehow are determined to be a part of the same ensemble. The acausal structure of this statistical modification of classical GR should make it apparent that our physics is simply not like that (deep down)! But the same point would apply to (2.1) itself if we took $\langle \hat{T}_{\mu\nu} \rangle$ as a classical expectation value over an ensemble of runs of a (quantum) experiment.

Of course, in quantum theory, there is a ready and standard reading of "expectation value" applicable to a single system: The sum of eigenvalues weighted by the amplitudes squared of the corresponding terms in the quantum state of that system in an individual run. While the Born Rule entails that the ensemble average will (probably) agree with that value, the quantity itself is well defined in terms of the state of the single system, unlike the classical case. Even so, we will see in Section 3.2 – in the context of the measurement problem – that there can be ambiguities in how we move between the classical reading of the expectation value and the quantum reading in analyses of quantum experiments.

How one constructs a quantum stress-energy tensor operator in QFT is a subtle business. But, once defined, it is indeed an operator that acts on states $|\psi\rangle$ of a material quantum system. As such, the states will obey the Schrödinger equation, with Hamiltonians describing both the dynamics of matter with itself, and with gravity:[3]

$$i\partial_t |\psi\rangle = \hat{H}_{\text{matter+gravity}} |\psi\rangle. \tag{2.2}$$

A system described by Eqs. (2.1–2.2) is often referred to as "semiclassical gravity" (SCG), and the Møller–Rosenfeld equation rechristened the "semiclassical Einstein" equation. But this usage hides an important ambiguity, which can (and does) lead to significant miscommunication. On the one hand, one might take Eqs. (2.1–2.2) as jointly comprising an *approximation* to the dynamics of a full solution to the problem of QG, perhaps along the lines of string theory or loop quantum gravity. On the other hand, they might be proposed as a *full solution* to the problem themselves: that is, where gravity is fundamentally classical, so that the quantum nature of matter entails a semiclassical theory. Let's take these two possibilities (and a third that will arise) in turn.

---

[3] Note in interpreting both (2.1) and (2.2) in terms of a common notion of time, in this Element we sweep under the carpet the "problem of time" familiar in QG research, without further comment.

View 1: SCG as a Mean-Field Description in Low-Energy Quantum Gravity The first reading holds that classical GR succeeds under ordinary circumstances because they reside in a regime in which it provides a good approximation to an as-yet unknown fundamental theory of QG (which need not be a final theory of physics). More particularly, models of GR are taken to be "mean-field" solutions of the unknown theory, and quantum perturbations around those solutions are taken to provide an effective field theory (EFT) for the underlying unknown theory: A ubiquitous implementation of this EFT is quantized linearized general relativity.[4] This picture is discussed and its many concrete applications described in, among other places, Burgess (2004) and Wallace (2022), both of who call it "low-energy quantum gravity" (LEQG).[5] Their point (also made by Crowther and Linnemann [2019]) is that such a theory is both empirically successful and constitutes a quantum theory of gravity in any reasonable sense: indeed, in just the sense that we have a quantum theory of the electromagnetic, weak, and strong interactions (viz., a UV-incomplete EFT of some higher energy physics).

In the LEQG framework, SCG amounts to the low-energy limit of the gravitational EFT, when quantum fluctuations in the gravitational field may be ignored. For instance, in the case of quantized linear gravity, Hartle and Horowitz (1981) derive (2.1) as the lowest order quantum matter corrections, in the large N limit of N gravitating quantum systems (a fact to which we will return). But one also expects SCG as a limit on more general grounds than that derivation: Sakharov's "induced gravity" program (Visser 2002) begins with the observation that (2.1) holds in the limit for *any* theory that dynamically couples a Lorentzian metric to a quantum field.

Within the approach described by Burgess and Wallace, SCG then solves the problem of incorporating quantum matter into classical

---

[4] In the applications that we consider, linearization occurs in a Minkowski background, so the perturbations can take the form of massless, spin-2, gravitons. See Huggett and Wüthrich (2020, Section 10) for further discussion. Note too that we leave open whether "mean field" is understood more literally or more analogously, depending on the nature of the underlying QG theory and the limit taken within that theory to obtain the EFT.

[5] Note that LEQG does not comprise all that might be termed "low-energy quantum gravity": For instance, perturbative quantum cosmology (or even nonperturbative, yet symmetry-reduced quantum cosmology in general).

dynamical spacetime theory, provided that we restrict attention to an appropriate regime in LEQG.[6] (Indeed, one could take SCG as showing that there was no real "problem" in the first place.) On this picture, given the strength of the gravitational interaction relative to others, one would expect in this regime leading order corrections to the expectation value of stress-energy, for a given state of matter, to come from quantum fluctuations of the matter field itself, and not from gravitational fluctuations (at least away from strong curvature). Then, even though one expects deviations from classical gravity in the long term because of the nonlinear character of (2.1), for short durations of time, it is sufficient to model such a system in terms of accumulating effects of back-reaction by matter corrections on the spacetime curvature, which would otherwise determine the left-hand side of the equation. Such a modeling program is known as "stochastic gravity," and has been successfully developed (see, e.g. Hu and Verdaguer [2008]). As noted by Wallace (2022, p. 39), stochastic gravity may be derived from LEQG, indicating no tension between the present view of SCG and the expectation that stochastic noise due to the quantum nature of matter influences the effective classical description of gravity.

What is crucial to this view is that gravity is understood as fundamentally quantum in nature, and only effectively treated as classical, that is, for the purposes of approximation. This approximation is summarized by the dictates of SCG, interpreted in terms of LEQG, perhaps improved with corrections from stochastic gravity.

View 2: SCG as a Candidate Fundamental Theory of Gravity

According to a second possible view, (2.1–2.2) constitutes an *alternative* to string theory, loop quantum gravity, and so on for a fundamental theory in its own right, not approximation. This view is not taken as a serious possibility by the QG community, yet there have long been efforts to rule it out definitively: Huggett and Callender (2001) critique theoretical arguments and also the experimental approach that we discuss

---

[6] And, we would add, an appropriate regime within the underlying QG theory that recovers LEQG in a suitable limit. This regime may not be quite identical to that picked out in LEQG – indeed, one would hope not, if QG research is to resolve outstanding problems in LEQG like the cosmological constant problem and black hole evaporation.

later. Likely, that this view is not taken seriously in part reflects the fact that it is beset by mathematical difficulties. Namely, it is not clear that any spacetime and quantum field could simultaneously satisfy both equations. One can define a QFT satisfying (2.2) in a given curved background spacetime (and perhaps solve the equations), but once one has, likely its stress-energy will not satisfy (2.1) in the background. One might then introduce a new background for which (2.1) is satisfied, but now (2.2) will likely not hold, and a new QFT must be defined. And so on. As we say, perhaps the equations can be solved simultaneously, perhaps the process described even converges on a solution (and perhaps merely approximately so), but it is far from sure.

Still, physicists are no strangers to mathematical difficulties in the course of theoretical research. Arguably, what is more *conceptually* troubling is the disunity involved in accepting that some parts of the world are classical while others are quantum even at the most fundamental level. Moreover, there are difficulties contemplating what even some basic physical models of such a theory would look like, beyond maybe the vacuum sector. For these reasons, in conjunction with the mathematical difficulties, it is perhaps not surprising that this view is given little credence by physicists. At the same time, it is considered worth eliminating as a live possibility.

Moreover, it is not even clear that Eqs. (2.1–2.2) adequately capture what is claimed: a meshing together of classical gravity and quantum matter, as we currently understand them. Recall in view 1 that corrections due to matter fluctuations, as studied in stochastic gravity, plausibly are relevant to the gravitational properties of a classically gravitating quantum system, so that it is a virtue of that view that LEQG recovers familiar stochastic gravity techniques. As just stated, view 2 categorically denies the role for any such corrections. The result is that there is rampant loss of physical information in coupling the material quantum system to gravity: After all, expectation values are insensitive to all higher-order correlators in the QFT.

View 3: SCG as A Mean-Field Description of X   Now, taking SCG as an approximation does not in itself commit one to the view that SCG is an approximation to *LEQG*; perhaps the low-energy approximation to a fundamental theory of QG is not LEQG, but some other theory, so that SCG is a "mean-field description" of that. View 3 is thus an epistemically careful deviation from view 1: SCG is a mean-field description but one stays uncommitted to the specific microstructure X. Now, there are few serious advocates for such alternatives, but some have been mooted. First, Carney et al. (2019) discuss a toy model in which a mesoscopic "ancilla" continuously weakly monitors the fluctuations in the microscopic quantum fields, feeding the inevitably noisy record of those fluctuations directly into a classical gravitational interaction channel.[7] More prominently, thermodynamic derivations of GR proposed by Jacobson (1995) and Padmanabhan (2014) render the degrees of freedom as collective ones; such derivations are often taken by these authors and the wider 'emergent gravity' community as suggestive that GR's degrees of freedom are therefore not to be quantized. (Though, we note, such a strong conclusion seems too quick, given that the quantization of collective degrees of freedom can sometimes result in a better approximation, as is arguably the case for sound waves whose quantization leads to phonons.) And a final non-LEQG viewpoint on SCG, which is often taken as expressing a similar view of spacetime geometry as collective degrees of freedom, is Sakharov's induced gravity (Visser (2002) for a review): SCG is seen as a consequence at lower energies for *any* sufficiently rich quantum field theory scenarios, just provided that one assumes a dynamical metric at lower energies. Of course, such results supporting the induced gravity program do not rule out SCG being an approximation to LEQG, but they do indicate a (conceivable) universality of SCG: that it could arise in a suitable limit in many conceivable theories.

As the range of examples illustrate, the point is not to advocate for a particular theory (contrast this with LEQG in view 1). Rather, the coexistence of the different examples demonstrates that one could, for the time being, remain *neutral* on the exact relationship between QG and

---

[7] A version of this model was first proposed and studied by Kafri, Milburn, and Taylor in a pair of articles (Kafri et al. 2014, 2015), and so is sometimes referred to as the KTM model. As elaborated in Altamirano et al. (2018), there are strong observational constraints on proposals for which gravity is fundamentally classical and meanwhile the pairwise gravitational interaction between macroscopic quantum sources is effectively Newtonian.

SCG. This would be to accept that SCG approximates some theory X, which in turn approximates fundamental QG (or even something wilder), but meanwhile to take no stance on whether X is LEQG, or some theory in which gravity is classical, or some alternative, considered or unconsidered.

In the discussion so far, introducing view 3 may seem unmotivated; what is the point of singling out this particular epistemic stance? The vast majority of theorists seem to adopt view 1, for instance. One theme through the remainder of this Element is that the differences across these three different views of SCG, given our current best physics, really matter to how we understand experiments in tabletop quantum gravity. Take for instance the Page and Geilker tabletop experiment to be presented in the next section. We will show how the extremely strong interpretation of the experimental results given by the original authors, in contrast to the typically deflationary attitude found in response to those results, reflects the difference between SCG view 1 and SCG view 3, where each is understood as opposed to SCG view 2. Getting clear on the Page and Geilker experiment in this manner will later pay dividends in clarifying the terms of the disagreement regarding the significance of the proposed GIE experiments.

3 Experimental Prelude: Quantum Probes in Two Traditions

In the previous section, we explored three ways of understanding SCG, in the context of ongoing theoretical research regarding the problem of QG. Here, we discuss two experiments in the history of tabletop quantum gravity. As we will stress, the standard presentations of what is (or is not) accomplished in performing each of these respective experiments would seem to represent two different traditions in experimentation within the discipline, or two different ways of hoping to provide increasingly better empirical traction on the same problem. Following this section we will, in part, consider the new class of experiments with gravcats by the lights of each of these traditions.

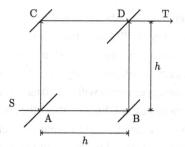

Figure 3.1 Schematic diagram of the COW neutron interferometry experiment

3.1 COW

Consider first the COW experiment (named after Colella, Overhauser (1974) and Werner (Colella et al. 1975)): Neutrons pass through a beam splitter and the different components of their wavefunction follow paths at different heights in the Earth's gravitational field, producing a relative phase shift between the components, which can be observed as interference on recombination. Our discussion of this scenario will largely be within NRQM, but it can also be modeled covariantly to little difference, as we shall briefly see. As with all the cases we discuss, the question of whether the effect is quantum is delicate.

The basic setup is shown in Figure 3.1, with the Earth understood to be at the bottom of the diagram: The neutron beam enters from the source S at left, is split at A, with the two components following lower and upper paths and recombining at D, with interference due to any relative phase of the components observed at T.[8] How do we now calculate the resulting shift?

Consider a nonrelativistic description first: Let us suppose that the neutron beam is well described by a plane wave, $\psi(r,t) = Ce^{-i(pr-Et)/\hbar}$ as dictated by the Schrödinger equation, so that the phase is $pr - ET$ in $\hbar = 1$ units (here r is the distance along a path). Then one might suppose that the effect is the result of CD and AB being at different heights in the

---

[8] The geometry of the actual experiment was a parallelogram, and the device could rotate about a horizontal axis in its plane to change the relative heights of the paths and measure the change in phase shift.

Earth's gravitational field, and hence corresponding to different gravitational potential energy (GPE) for each component along those path
segments. But this would be a mistake. First, neutron energy is of course
conserved, with changes in GPE canceling with changes in kinetic energy
(KE). (Note that in the experiments with gravcats discussed later, the KE
will be zero, and differences in GPE can affect the phase through the Et
component.) Of course, a neutron will decelerate along AC and hence
take longer to traverse CD than AB, and so the lower path ABD represents less time than the upper path ACD, even though the energy is the
same along both. Does this lead to a phase shift of $E \cdot (t_{ACD} - t_{ABD})$? No.
Because the part of the wave reaching D along the upper path ACD takes
an extra $t_{ACD} - t_{ABD}$ to get there, it must have left S a time $t_{ACD} - t_{ABD}$
earlier than the wave along lower path, in order for them to arrive simultaneously. Hence they would have been out of phase on emission by exactly
$-E \cdot (t_{ACD} - t_{ABD})$, exactly canceling the difference picked up around
the path.[9] In other words, the Et part of the phase contributes no relative
phase, and the whole observed effect is due to the pr part of the phase.
Before we go on to calculate this quantity, note that when considering the
deceleration of neutrons due to gravity, we have treated them as classical
particles. Indeed, the entire calculation of pr will proceed in this way,
with implications for the quantum nature of the phenomenon.

The original CO (Overhauser and Colella 1974) prediction is based
then on lower momentum along CD than AB, because of gravitational
deceleration along AC. Basic mechanics tells us that a particle of mass m
and initial vertical speed u will have a final speed $\sqrt{u^2 - 2gh}$ after travel-
ing a distance h; or to first order in g, a speed of $u - gh/u$. Assuming that
the reflection is perfect, the phase change along CD is thus $m(u - gh/u)h$,
while that along AB is simply muh since no deceleration is involved. The
momenta along the vertical legs should be the same, and so CO predict a
relative phase shift of mgh^2/u, which is indeed observed.[10]

---

[9] This result is general. If two parts of a coherent wave travel along paths to a target
   T in different times, the resulting phase shift is exactly canceled by the time difference in emission. Mannheim (1998) nicely explains how this effect is necessary for
   understanding the familiar two-slit experiment.

[10] In fact the observed value was around 10 percent lower. The largest correction comes
   from deformation of the crystalline structure of the apparatus, which is carved from a
   single crystal of silicon. Once this is accounted for, other effects can be measured by

However, since CO use classical particle considerations to calculate the motion of the neutron beam, they should do so consistently, for instance, taking into account that paths will not be horizontal but parabolic as the neutrons fall during their flight: They reach the plates not at B and D, but slightly lower. This perturbation affects the time, magnitude, and direction of the paths, and at first order so must be accounted for. CO reference this effect, but the analysis of Mannheim (1998) shows that they do not correctly compute it: Indeed, he shows that there is *no* relative phase along the closed part of the paths! Instead, the particle considerations that CO invoke, when applied consistently, reveal a different source for the observed interference – which happens to agree with their original computation! (With hindsight, the coincidence is perhaps not a great surprise given that mgh^2/u is the natural dimensionless quantity in the problem.)[11]

Though the details of Mannheim's calculation are somewhat off the main line of this Element, the bottom line is, first, that there is no phase difference at all between a point on the splitter and the recombining screen along the two paths. However, second, a point on the wavefront at the splitter will not reach the same point on the recombining screen; rather, wavefront points that are slightly offset on the splitter reach the same point of recombination. Finally, since the splitter is not perpendicular to the neutron beam, such wavefront points will travel different distances to the splitter, and so already be out of phase at the splitter, *completely accounting for the effect*!

One might thus conclude that the effect is not gravitational at all! But this would be too hasty: Although the phase changes along the paths cancel, their explanations involve two quite different combinations of speed and trajectory shifts, and the Earth's gravity is the cause in both. Indeed,

---

further deviations from the prediction, including that due to the Earth's rotation. See Greenberger and Overhauser (1980).

[11] Greenberger and Overhauser (1979) show that under certain assumptions, the effect of parabolic motion is (to lowest order) *numerically* equivalent to integrating the potential difference along the paths, and thereby also derive the same result. However, they do not take into account all the factors that Mannheim does: especially the change in momentum and reflection angle caused by parabolic motion. So they do not establish the full equivalence of their method and Mannheim's analysis. We also note that Berry (1982) provides an exact calculation of a different quantum effect involving "neutron rainbows" in the Earth's gravity.

in the actual experiment, when the apparatus is rotated about a horizontal axis through its plane, so that the "vertical" paths are no longer vertical, and the horizontal paths change their relative heights, the phase changes no longer cancel, and the observed relative phase changes.

As an aside to our main topic, we note that observing a phase shift mgh^2/u amounts to a measurement of the neutron mass, since the size of the experiment and the speed of the neutron are known. This is remarkable, because in the vicinity of the equivalence principle is the idea that it is impossible to measure the mass of a body using purely gravitational effects – but that is exactly what is accomplished by the COW experiment.[12] Indeed, one might therefore worry that the experiment reveals an incompatibility between the equivalence principle and NRQM. However, reflection on the equation of motion, the Schrödinger equation, of course reveals the identity of gravitational and inertial mass that constitutes the Newtonian equivalence principle: The same mass m appears in the kinetic term that governs inertial motion, and in the potential term that determines the gravitational force. (The same points hold, mutatis mutandis, in a covariant analysis (Mannheim 1998, p. 57).)[13]

What though does the experiment show about the relation between Newtonian gravity and NRQM? Clearly the effect is quantum in the sense that in the $\hbar \to 0$ limit the neutron is purely classical, and has no phase at all. Additionally, in Mannheim's analysis, the effect depends on the finite spatial (and temporal) spread of the wavefunction. However, our question regards the nature of the *interaction* of gravity with quantum matter, the neutrons in this case. So a more relevant consideration is that the GPE has no direct effect on the phase, as one might have naively thought: As we discussed, the energy is conserved (and the time from source to target is irrelevant) so we find no Et component of the phase. Moreover, the calculations of p and r, which do contribute to the phase, depend on purely classical properties of the neutrons: their trajectories determined by Newtonian mechanics in a constant gravitational field. So in that regard, gravity and matter are related purely classically in the experiment. (Note that all these considerations – except the finite spatial extent of the wavefunction – apply equally to CO's analysis.)

---

[12] This point was made in Okon and Callender (2011).

[13] Brown (1996) considers other implications of the experiment, especially for the action–reaction principle.

As we mentioned earlier, we have focused on an NRQM model of the experiment, in order to simplify a fairly complex situation. However, Mannheim also outlines a covariant model, which it is worth briefly discussing to see that it produces the same result, and also as a warm-up for later calculations. First, for the metric describing the Earth's gravitational field we use the weak-field approximation to the Schwarzschild solution to the Einstein field equation:

$$ds^2 = (1 + 2V(\vec{r})/c^2)dt^2 - d\vec{r}^2, \tag{3.1}$$

where $V(\vec{r})$ is the Newtonian potential, in this case $-gx$ when the Earth is the source of the field. Timelike geodesics then satisfy $\ddot{y} = \ddot{z} = 0$, $\ddot{x} = g$, so that in this approximation neutrons will follow the same paths that we computed in the Newtonian analysis. Moreover, treating the neutrons as excitations of a Klein–Gordon quantum field reveals that the phase change is again dependent on the momentum along each path. Hence overall, Mannheim's Newtonian calculation of neutron interference applies equally to the covariant treatment.

That said, the covariant situation is different, insofar as the effect also has an entirely classical realization, unlike a Newtonian analysis. This is because covariantly light will follow null geodesics, and not the straight and vertical paths expected in a Newtonian treatment. Remarkably, the very same considerations that apply to massive particles apply to massless classical waves: Interference is predicted for a classical light beam COW experiment, for exactly the same reasons as for the quantum neutron experiment. Thus, one can envision a gravitational Michelson–Morley experiment to observe the bending of light in the Earth's gravitational field![14] Thus, there is a purely classical realization of the effect covariantly, though not in the Newtonian treatment.

Since one expects the analog of the COW experiment even in this classical system, it seems that the COW experiment's credentials as an observation of the quantum nature of gravity are at best weak. But looking at things in this way misses the main point of the experiment.

---

[14] We thank Professor Mannheim for discussions of these matters. He reports that an apparatus with arms 1 km long would be capable of measuring the effect: In stronger fields, of course, such an apparatus would only need shorter arms to measure the bending of light.

Instead, one is interested in treating the gravitational interaction channel as opaque: to be studied with increasingly sophisticated quantum probes, in terms of the effects registered in those probes. In doing so, one hopes that physicists might develop good experimental *control* over the gravitational behaviors of material quantum systems that are increasingly poorly described as classical, providing new knowledge (a "causal manipulation" kind of knowledge) that is relevant in the course of theoretical research on the problem of QG. Here, for instance, one may study the dynamics of individual quantum probes isolated from their (known) quantum environments, but which are nonetheless subject to gravity for the duration. Later, we will evaluate GIE experiments similarly, as expanding the range of experimental quantum probes of gravity. So, simply to emphasize this sense of continuity between COW and GIE, we speak of a "control *tradition*."

3.2 Page and Geilker

The second experiment is that conducted by Page and Geilker (1981), and reported as "indirect evidence for quantum gravity." So begins, in contrast with what we have just said about COW, the tradition of experiments in tabletop quantum gravity that claim to search for increasingly sophisticated *witnesses* of the underlying quantum nature of gravity.

The Page and Geilker experiment is a modification of the famous Cavendish experiment. In the original, two smaller test masses are placed on the ends of a hanging torsion balance, and a larger source mass is placed close to each, but on opposite sides of each end: The gravitational attraction between the test and source masses is thereby measured by the deflection of the balance. If the weights are placed in, say, position A in Figure 3.2, then the balance will deflect counterclockwise. In the authors' modification, quantum mechanical radioactive decay is used to randomize between two classical configurations: the weights at A exerting a counterclockwise torque, and the weights at B a clockwise torque. Since radioactive decay is a quantum process, Page and Geilker reason that in each run the weights will be in a superposition of classical positions A and B. The *expected* position of the weights in such a state is between the two positions, co-located with the test masses, and so

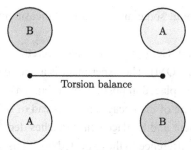

Figure 3.2 Schematic diagram of the quantum Cavendish experiment viewed from above. A and B represent the two possible locations of the weights; on each run, the actual position is decided by the reading of a Geiger counter

SCG – which depends on the expected distribution of matter – predicts that no net gravitational force is exerted on those test masses, and the torsion should vanish in every run.

Of course, it was no shock that when Page and Geilker performed the experiment, they observed instead that the balance always deflected, with equal frequency in each direction: Although SCG predicts the correct expected displacement, it is wildly incorrect with respect to the individual runs. As Ballentine (1982) dryly puts it, "a less surprising experimental result has scarcely, if ever, been published." (The observed result is of course compatible with an entirely different setup, one in which the source weights are *randomly* prepared in each trial in just one of the two classical configurations. So if the experiment is to rule out SCG, it is crucial that the choice of position is determined quantum mechanically.)

In the first place, Page and Geilker intend that the experiment serves as a refutation of SCG, construed as a candidate fundamental theory of gravity, as in view 2. As such, the result is of course unsurprising – we do not expect macroscopic source mass positions to coherently superpose! (Thinking of our earlier discussion of the Møller–Rosenfeld equation specifically, we expect the torsion in any one run to depend on the position of the sources in that run, not their average position over a series of runs.) However, caution is merited. As the experiment depends on the outcomes of individual runs rather than a statistical average for an ensemble, one simply cannot interpret the experiment as a test of SCG without some answer to the "measurement problem" in mind.

Indeed, Page and Geilker assume an Everettian interpretation of quantum mechanics (QM): The experiment is a massively open system, since the sources are placed in position by a (human) experimenter who observes the outcome of the decay. Thus, instead of being in superposition, the A-placement and B-placement branches decohere. Such decoherence makes no difference to the expected positions of the weights, so the SCG prediction is still no deflection. The observed deflections are understood instead as a random (with respect to the Born Rule) sampling of branches over an ensemble of experiments, in each of which the deflection is toward the position of the weights in that branch. Granted then, the experiment refutes SCG as a candidate fundamental theory, given an Everett interpretation.

Of course, standard interpretations of quantum mechanics predict the same outcome if SCG fails. As Wallace (2022, Section 5) points out, if an interpretation allows only unitary dynamics, then it will appeal to decoherence in the way just described. While if an interpretation invokes some kind of collapse, real or effective, it will occur when the experimenter measures whether a decay occurred as part of the procedure, "collapsing" the system into one of the branches. Thus, no one would plausibly expect a macroscopic superposition of the source masses in the experiment, so at the macroscopic scales relevant here, classical descriptions of (fundamentally) quantum physics suffice to capture, *in each run*, the gravitational dynamics of the experiment.

However, for this very reason, interpretations that go beyond unitary dynamics may also predict the same outcomes even if view 2 of SCG is correct. For example, since collapse interpretations such as that of Ghirardi, Rimini and Weber (1986) entail a collapse once the decay (or its absence) is observed by the experimenter, they predict that the source masses are never in superposition: So, the expected mass distribution is always either in position A or in position B. Similarly, a hidden variable theory, such as Bohm's, could trace a definite configuration to some uncontrollable – hence effectively random – epistemic uncertainty in the initial state, leading to "effective collapse" (Dürr et al. 1992, Section 3.2). In other words, the implications of Page and Geilker's experiment are sensitive to the interpretation of QM: According to the Everett interpretation, it does serve to rule out SCG, while for collapse or hidden variables it may not do so. One would like to close such loopholes; the experiments with gravcats to be discussed later would achieve this goal (and maybe rather more).[15]

Page and Geilker not only take – according to their Everettian lights – the experiment to refute view 2 of SCG, they also take it to provide indirect evidence for the quantum nature of gravity. This point can be readily understood in terms of Bayesian updating of probabilistic degrees of belief ("credences"), a framework on which we will draw at several points. It is a trivial theorem of probability that the probability of a hypothesis H, *conditional* on a possible piece of evidence E is given by $p(H|E) = p(E|H)p(H) \div p(E)$. Bayesian updating proposes that if E becomes known to be true – if, that is, an experiment or observation produces outcome E – then, on pain of irrationality, one's credences should change so that $p_{new}(H) = p(H|E)$, the "posterior (to the updating) credence."[16] Now suppose that there are a number of hypotheses H_i, all but one of which, $\bar{H}$, entail E; while $\bar{H}$ entails $\neg E$. Then suppose that outcome E is observed, so that $p_{new}(\bar{H}) = 0$. Because probability is conserved, $p(\bar{H})$ must be redistributed over the remaining hypotheses; from the theorem (since $p(E|H_i) = 1$), this will be in proportion to $p(H_i)$, the "prior" for H_i. The standard picture is a quantity of sand evenly spread over a collection of boxes of different sizes; if one box has to be emptied, its contents are spread over the other boxes, in proportion to their sizes.

This situation, roughly speaking, describes the Page and Geilker experiment. If it rules out view 2, then the prior probability gets spread across the remaining alternatives including view 1: The quantum nature of gravity is confirmed in the sense that the credence for SCG as a mean-field description in LEQG increases.

As we explained, whether this conclusion holds depends at least in part on attitudes toward quantum measurement, and so whether view 2

---

[15] However, we concur with Wallace (2022) that physicists who claim that standard unitary quantum dynamics is complete really *should* accept the Page and Geilker experiment as experimentally refuting view 2.

[16] There is a nearly endless literature on credence as probability and Bayesian updating: for example, see references in Lin (2022).

is refuted.[17] But even supposing that it is, how seriously one takes the refutation of view 2 to count as meaningful evidence for the quantum nature of gravity – say, substantive shifts between prior and posterior credences – also depends on downplaying the possibility of some alternative to LEQG producing SCG in a low-energy limit, namely view 3. On this understanding of SCG, the observed classicality of the gravitational interaction within the individual runs of their experiment would not necessarily mean that gravity is fundamentally quantum. Since the prior for view 2 is spread across views 1 and 3 in proportion to their priors, then the degree to which view 1 is confirmed depends both on one's prior for view 2 (how likely it is that SCG is the final word on QG), and the relative priors for views 1 and 3 (how much of the available probability accrues to each). One can then understand the muted reactions of physicists to Page and Geilker's paper. Whether one already has a low prior for view 2, or has comparable priors for views 1 and view 3, or adopts a collapse interpretation of QM, not much is to be learned from performing the experiment.

4 Gravitationally Induced Entanglement Experiments

The COW experiment, while showing that gravitational fields affect quantum matter, considers only the field of a classical source, the Earth. The Page and Geilker experiment claims to involve the gravitational fields of sources in quantum superposition (though decohered), but its interpretation as evidence of QG involves loaded assumptions. No wonder then, that many have looked for more clear-cut demonstrations of the "quantum nature of gravity." And so we turn to the main object of this Element: Recently proposed experiments that arguably can demonstrate the quantum nature of gravity, and thereby present as a new frontier in the witness tradition that began with Page and Geilker. In Section 4.2, we will give a first account of these new "tabletop" experiments. They centrally feature a "gravitational Schrödinger cat," or "gravcat" (as the neologism mentioned earlier has it) – an uncharged object large enough to

---

[17] A more complex analysis that continues to make use of the discussed formalism would consider more elaborately specified hypotheses, which basically amount to conjunctions of interpretations of quantum mechanics and claims about SCG.

exert a gravitational force greater than any van der Waals forces, but small enough that it can be placed and preserved in a quantum superposition for long enough to interact with another gravcat and have their entanglement measured. It has recently been proposed by Bose et al. (2017) and Marletto and Vedral (2017a) that a pair of gravcats could be used to probe the quantum nature of gravity, in experiments that could in principle, and perhaps in practice within the next decade or so, be carried out on a lab bench (perhaps using 10^{-14} kg diamonds).

We briefly note that the phenomenon studied in these experiments is not the first within our grasp that purportedly relies on the quantum nature of gravity, independently of one's favored interpretation of QM. The explanation of cosmic microwave background (CMB) structure in terms of fluctuations in the inflaton field in the early universe requires gravity to be a quantum field (as Wallace (2022, Section 6) discusses). However, drawing conclusions about the quantum nature of gravity requires considerably more theoretical assumptions than for the new experiments: not only the standard theory of inflation, but also that stochastic gravity cannot equally explain the CMB data (as proposed by Roura and Verdaguer [2008]). So the new experiments are an advance both by exploring a new regime, and by their relative theoretical neutrality.

4.1 What Makes a Witness

Describing such a "gravitationally induced entanglement" (GIE)[18] experiment, Marletto and Vedral write: "the entanglement between the positional degrees of freedom of the masses is an *indirect* witness of the quantization of the gravitational field" (our emphasis): Such experiments are distinct from (and far easier than) "direct" observations of the quantum nature of gravity, for example, from the observation of individual gravitons or their effects, perhaps some gravitational analog of the photoelectric effect.[19]

---

[18] Names are not entirely settled. One also hears "gravitationally mediated entanglement," but we have avoided this term as it may not sound sufficiently neutral with respect to the paradigms discussed in this Element.

[19] Observing a strict analog to photoelectric effect is fanciful: for the $n = 2$ to $n = 1$ hydrogen transition, the electrical potential energy is 10eV, but the gravitational potential energy is 10^{-38}eV, so while the former is readily observable, the latter is not (to say

Let us remark briefly on the "direct/indirect" distinction, as the GIE literature seems to be at cross purposes on this point. Namely, Marletto and Vedral's classification of the experiment as *in*direct (also Carney et al. [2019]) is at odds with the discussion found in the supplementary materials of Bose et al. (2017), where one reads: "it is fair to say that there are no feasible ideas yet to test [QG]'s quantum coherent behavior directly in a laboratory experiment. Here, we introduce an idea for such a test...." Likely, this clash is terminological. For instance, Christodoulou and Rovelli (2019, p. 65) cite both as supporting the same conclusion: "As emphasised by Bose et al. and by Marletto and Vedral, the main reason for the interest of the experiment is precisely to provide *direct* evidence that gravity is quantised" (our emphasis). So perhaps there is consistency across these claims, but some competing intuitions about how to draw a direct/indirect distinction, with the notion of "witness" absorbing tension between direct evidence and a lingering sense that the results of the GIE experiments would still somehow be "indirect." One philosophical inclination is, of course, to attempt precisification: definitions of the terms that are consonant with familiar uses, or intuitions of "direct/indirect" in other physics contexts (e.g. Franklin [2017]; Elder [2022]).

However, after our unsuccessful attempts to obtain a useful clarification of the terms, it is our conclusion that there is simply insufficient "data" on how they are used in the context of GIE. Meanwhile, as we aim to show, the most interesting philosophical question concerns the nature and possibility of disagreement over the significance of a positive GIE result, and that disagreement does not at all turn on the question of a direct/indirect distinction. For instance, none of those quoted question that it would "witness" the quantum nature of gravity; and none of those disputing the latter would be placated by being told that the witness was not "direct," but merely "indirect." For this reason, to clean up actors' terms, we will here set aside the "direct/indirect" language, and assert for the sake of moving forward: All players mentioned intend the experiment as a "witness" of the quantum nature of gravity. We maintain that

---

the least!). More practical – but still impossible by many orders of magnitude – is the observation of gravitationally induced decoherence of entangled systems (see Carney et al. [2019, Section 3.2]), though arguably this too would be "indirect."

mere stipulation of this kind does not diminish the task at hand, namely, of evaluating the stakes of the new experiment from within the witness tradition; as we shall soon see, there is a real question of whether it would be a witness at all.

As for "witness," the term itself comes from an analogy with the term "entanglement witness" from quantum information theory: An observable acting on a given tensor-product system, whose expectation value for an entangled state of the system is incompatible with the range of expectation values for a separable state (Vedral 2006). Measuring such a value in the laboratory then constitutes a "witness" of entanglement, in contrast to a measurement of Bell-inequality violating correlations (the latter closer to an observation of entanglement itself). Analogously, in a GIE experiment instead of observing quantum behavior of gravity itself, observed entanglement (itself perhaps observed via an entanglement witness) is incompatible with purely classical gravity. In both cases, the broad idea is to measure a proxy for the feature of interest. Still, we prefer an even broader understanding of the term, divorced from technical notions: empirically accessing a specific (theory-laden) feature of the world. It is this totally nontechnical use of the term, which licensed us previously to identify the Page and Geilker experiment as the beginning of the "witness" tradition in tabletop quantum gravity experiments.

We will return to the question of what to make of GIE experiments beyond witnessing in Section 7. For now, with the previous discussion in mind, let us turn to a first, naive pass at the experiment itself, to get the basic ideas in hand.

4.2 A Naive Account of the Experiments

In the experiment, two identical gravcats are prepared in position superpositions along the x-axis. Let us say that the first gravcat is the sum of Gaussians located at $-D \pm \Delta$, while the second is a sum of Gaussians located at $+D \pm \Delta$; let us set $D \gtrsim \Delta$. Given their macroscopic size, we can take the gravcats to have no initial motion along the x-axis. Moreover, provided that the duration of the experiment is short, the gravcats will not have time to accelerate due to mutual gravitational attraction; they are effectively stationary. The arrangement is shown

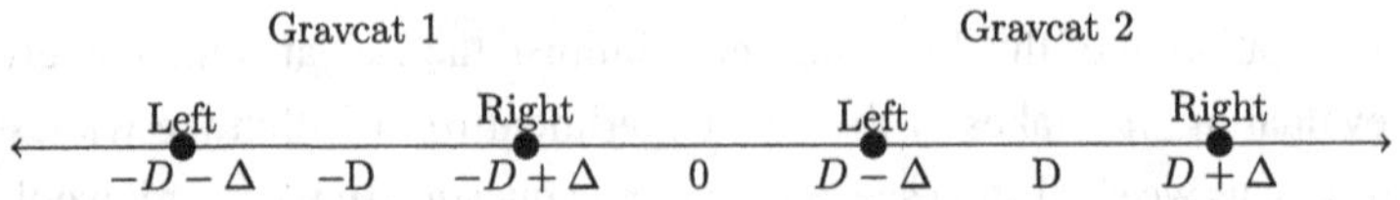

Gravcat 1   Gravcat 2

Figure 4.1 The two gravcat arrangement (the central packets are closer than shown)

in Figure 4.1, and is described by the following state (ignoring overall normalization):

$$\Psi(0) = (|L\rangle + |R\rangle) \otimes (|L\rangle + |R\rangle)$$
$$= |L\rangle \otimes |L\rangle + |L\rangle \otimes |R\rangle + |R\rangle \otimes |L\rangle + |R\rangle \otimes |R\rangle, \quad (4.1)$$

with $|L\rangle$ representing the left wavepacket of gravcat 1 in the first tensor product slot, and the left wavepacket of gravcat 2 in the second slot, and so on.

Because the distances between the packets of the two gravcats are different, different terms in (4.1) will have different gravitational potential energy (GPE), and so will develop relative phases.[20] For example, for $|L\rangle \otimes |L\rangle$ the GPE will be $Gm^2/2D$, while for $|L\rangle \otimes |R\rangle$, it will be $Gm^2/(2D + 2\Delta)$. Since the GPE is the only contribution to the energy (kinetic energy is zero), these different potentials will produce a different phase in each term according to the Schrödinger equation: $\Phi(t) = e^{-iEt}\Phi(0)$ for energy eigenstates such as these (recalling that we set $\hbar = 1$). By linearity then:

$$\Psi(t) = e^{\frac{-iGm^2t}{2D}}|L\rangle \otimes |L\rangle + e^{\frac{-iGm^2t}{2D+2\Delta}}|L\rangle \otimes |R\rangle +$$
$$e^{\frac{-iGm^2t}{2D-2\Delta}}|R\rangle \otimes |L\rangle + e^{\frac{-iGm^2t}{2D}}|R\rangle \otimes |R\rangle. \quad (4.2)$$

Or, since $D \approx \Delta$, for short times ($t \ll 2D/Gm^2$)

$$\Psi(t) \approx |L\rangle \otimes |L\rangle + |L\rangle \otimes |R\rangle + e^{\frac{-iGm^2t}{\delta}}|R\rangle \otimes |L\rangle + |R\rangle \otimes |R\rangle, \quad (4.3)$$

with $\delta = 2(D - \Delta)$ the separation of the closest pair in (4.5). This wavefunction does not factorize except for special values of t (and neither

---

[20] In this initial presentation of the experiment, we treat gravity as Newtonian, but we emphasize that Bose et al. (2017), whom we follow here, argue that the effect should be understood in terms of coherent states of a quantized relativistic field. We will discuss this point in Section 5.2.2.

does the un-approximated wavefunction (4.2)), and so the mutual gravitational attraction "induces" entanglement between the gravcats. It is this entanglement that is claimed to provide an indirect witness of the quantum nature of gravity.[21] (In contrast to such an experiment, note that the COW experiment involves "gravitationally induced *interference*.") All of this is still beyond current technology, but not so far beyond that experimentalists aren't attempting to develop existing techniques to make such a measurement, or one along similar lines.[22]

Now, given our description of the experiment, the claim that observing such entanglement witnesses the quantum nature of gravity should seem puzzling: After all, it appears that we simply appealed to classical gravity, just as in the COW experiment (whether by their lights or Mannheim's). Yet, Colella et al. (1975) did not claim that the neutron interference which was observed provided a witness of the "quantum nature" of gravity[23] – indeed, we even suggested that the experiment as self-described ultimately represents work in a different tradition of experiment, one that

---

[21] In a little more detail, Bose et al. (2017) propose that the gravcats have spin. Prior to the experiment, they are in a state

$$|X_1\rangle(|\uparrow\rangle + |\downarrow\rangle) \otimes |X_2\rangle(|\uparrow\rangle + |\downarrow\rangle), \tag{4.4}$$

where X_1 and X_2 are the initial positions of the two gravcats, and $|\uparrow\rangle$ and $|\downarrow\rangle$ positive and negative x-spin states. To prepare the state (4.1), the gravcats are passed through a Stern–Gerlach device oriented along the x-axis, to produce

$$(|L\uparrow\rangle + |R\downarrow\rangle) \otimes (|L\uparrow> + |R\downarrow\rangle) = |L\uparrow\rangle \otimes |L\uparrow\rangle + |L\uparrow\rangle \otimes |R\downarrow\rangle$$
$$+ |R\downarrow\rangle \otimes |L\uparrow\rangle + |R\downarrow\rangle \otimes |R\downarrow\rangle. \tag{4.5}$$

After gravitational entanglement is produced, the gravcats are passed back through the Stern–Gerlach device to yield

$$|X_1\uparrow\rangle \otimes |X_2\uparrow\rangle + |X_1\uparrow\rangle \otimes |X_2\downarrow\rangle + e^{\frac{-iGm^2 t}{\delta}}|X_1\downarrow\rangle$$
$$\otimes |X_2\uparrow\rangle + |X_1\downarrow\rangle \otimes |X_2\downarrow\rangle, \tag{4.6}$$

using the same approximation as (4.3). This state exhibits spin entanglement between the gravcats, which could be observed by measurements of spin-correlations between the particles that violate Bell-type inequalities. Note that the gravcat interaction itself does not involve their spins, only their mutual gravitational attraction.

[22] For recent efforts in this direction, see, for instance, the conference rationale and program of the following ICTP-SAIFR event: www.ictp-saifr.org/qgem2021/ (last checked on December 15, 2022).

[23] Though Greenberger and Overhauser (1980, p. 76) make a very weak claim in this direction.

instead emphasizes *control*. How is the present appeal to classical gravity relevantly different? Of course, in the COW experiment, the source
of the field was considered as classical, namely the Earth; while in the
GIE experiment it is quantum, the gravcats themselves. But that is a point
about the gravcats, not gravity. (The same, of course, can be said of the
sources in the Page and Geilker experiment, noting that gravcats remain
in coherent superposition.)

Unpacking this issue is a major task of the remainder of the Element.
We will see how different theoretical starting points can lead one to
different conclusions. For instance, the naive account tacitly assumes a
bipartite state of two gravcats, acting directly on each other, through an
interaction Hamiltonian

$$\hat{H} = \frac{Gm^2}{|\hat{x}_1 - \hat{x}_2|}, \tag{4.7}$$

in which gravity appears as a classical potential (except, of course, it is,
formally, an operator).[24] But the claim that the experiment would witness the quantum nature of gravity requires that it is something about
the quantum nature of whatever the classical potential in (4.7) actually
describes that ultimately explains the predicted outcome. What we will
now work out explicitly is that whether or not one takes this view –
namely, whether the ultimate explanation for the predicted outcome of
the experiment concerns facts about an underlying gravitational field, or
(just) the classical potential, fashioned into an operator – amounts to a
metatheoretical choice of modeling framework, or "paradigm" in a "lite"
sense (compared to various stronger senses also found in the literature):
No rational argument alone can make the proponent of one paradigm
swap to the other (Section 5.3).

To foreshadow: We will present two paradigms relative to which one
can understand the GIE experiments. The first paradigm starts from principles of our best current physics to vindicate the naive account: That is,
it turns out not to be so naive after all. We shall see that there is at least

---

[24] Of course, in our scenario, the kinetic term is ignored, because the particles are stationary. Indeed, no contributions to the Hamiltonian that depend only on the state of just
one particle – for instance, a potential in an external field – will affect entanglement,
since they act only on one part of the tensor product. Only a term that depends on the
state of both particles – a mutual potential – will produce entanglement.

one sense in which a nonlocal interaction may be seen to fall directly out of the fundamental field dynamics of GR, within the context of these GIE experiments. In this paradigm then, gravcats do not witness the quantum nature of gravity. The claim to witness comes within a second paradigm, spurred by nearly 200 years of physics, including relativity, telling us that bodies do not act at a distance on one another, but rather their interactions are mediated by fields (cf. Hesse [2005]). Exactly how to understand the gravitational field is of course equivocal, and indeed here it will mean different things: In some contexts it means the metric field of GR, in others the spin-2 massless field that appears in the linearization of GR on a Minkowski background. But at the simplest level, to treat gravity as itself a dynamical system responsible for mediating between bodies is to model the experiments with gravcats as tripartite, rather than bipartite (as in the naive model), with the field joining the two gravcats in the labeling of the *full* quantum state.

However, before discussing further how claims that gravcats witness the quantum nature of gravity are dependent on the paradigm adhered to, in the next subsection we will briefly rehearse an uncontroversial yet significant consequence of the GIE experiment: namely, that it would rule out semiclassical gravity, as understood on view 2 in Section 2, in such a way as goes further than the QM interpretation-dependent refutation claimed by Page and Geilker.

4.3 Ruling Out Semiclassical Gravity

Quite simply, the observation of post-experiment gravcat entanglement would be in direct experimental conflict with semiclassical gravity, defined by Eqs. (2.1–2.2), taken as a fundamental theory of nature, because such an interaction cannot result in entanglement. (As far as we are aware, the only other phenomena purporting to show the same thing are the two that have already been flagged: the controversial Page and Geilker experiment and the cosmological theory-laden explanation mentioned earlier of CMB structure by fluctuations in an inflaton field.) Let us explain.

To simplify matters, we work in the Newtonian limit of Eq. (2.1), in which the stress-energy of each gravcat is merely its mass density, $m|\psi(x)|^2$; for each gravcat, a mass density zero almost everywhere, but with equal small Gaussian peaks at the locations of the two wavepackets. The effect of such a distribution is (Anastopoulos and Hu 2014, Section 1.2)[25] to introduce a potential into the Hamiltonian of the form:

$$\hat{H}_{\text{gravity}} = -Gm^2 \int dx' \frac{|\psi(x)|^2}{|x - x'|}. \tag{4.8}$$

Inserting this term into Eq. (2.2) yields what is known as the "Newton–Schrödinger equation." This approach has often been criticized, and while not plausible, arguably has not been refuted empirically; hence it remains to date as a possible theory of quantum matter and classical gravity. However, we can readily see that witnessing entanglement in the proposed way would amount to direct empirical evidence against the theory.

Since gravcats only interact gravitationally, $\hat{H}_{\text{matter}}$ is merely the kinetic energy, which vanishes for stationary particles. Hence, assuming that the experiment does not last long enough for an appreciable change in velocity, Eq. (2.2) will only introduce phases according to this gravitational potential, which we can see will not produce entanglement. Suppose that the potential, implicit in (4.8), in which the left packet of the first gravcat sits is V_{L1}, then for an initial state $|L\rangle$ the time-dependent Schrödinger equation dictates that the state is given by:

$$\psi_{L1} = e^{-iV_{L1}t}|L\rangle, \tag{4.9}$$

and similarly for the other packets. Then, by linearity, if the initial state is again (4.1), then the Schrödinger equation yields as the time-dependent state:

$$\begin{aligned} \Psi(t) &= e^{-iV_{L1}t}|L\rangle \otimes e^{-iV_{L2}t}|L\rangle + e^{-iV_{L1}t}|L\rangle \otimes e^{-iV_{R2}t}|R\rangle \\ &\quad + e^{-iV_{R1}t}|R\rangle \otimes e^{-iV_{L2}t}|L\rangle + e^{-iV_{R1}t}|R\rangle \otimes e^{-iV_{R2}t}|R\rangle \\ &= (e^{-iV_{L1}t}|L\rangle + e^{-iV_{R1}t}|R\rangle) \otimes (e^{-iV_{L2}t}|L\rangle + e^{-iV_{R2}t}|R\rangle), \end{aligned} \tag{4.10}$$

---

[25] They take Eq. (2.1) as effective; here we are taking it as fundamental, and thus their conceptual critique of the derivation does not apply. Note that Hu and Verdaguer (2020, Section 1.3) show that the Newton–Schrödinger equation does *not* arise in the combined weak-field limit of GR and the nonrelativistic limit of quantum field theory. Instead, what arises is the equation discussed in the next subsection under the label of "Newtonian Model."

so that the wavefunction factorizes, and there is no entanglement according to SCG.

It's easy to understand why: Because it is determined by the *expected* matter distribution, the gravitational field is the *same* in each of the four terms; thus, in any term, the potential acting on any gravcat depends only on its position, not that of the other gravcat; so the $|L\rangle$ (and $|R\rangle$) state of each gravcat picks up the same phase in any term in which it appears, as can be seen in (4.10). But then it is a simple mathematical fact that the state remains factorizable.

Hence the tabletop experiments offer an unequivocal empirical test of the quantum nature of gravity insofar as they rule out a classical treatment in the form of SCG, *if* gravcat entanglement is observed as hypothesized. According to LEQG and view 1, this failure is quite understandable; as Anastopoulos and Hu (2014, p. 5) note (and we mentioned earlier), SCG is a good effective theory in the limit of many bodies, and the gravcats are simply outside that regime. (It's worth noting that SCG fails to explain CMB structure for an entirely different reason, namely that contributions from quantum fluctuations must be included as corrections to the mean-field approximation in order to match the empirical data.)

That said, a real puzzle remains about the enthusiasm of the community for carrying out the experiment, given that it will require considerable effort and expense. After all, it is almost universally believed in the community that view 2 of SCG is in fact false (not withstanding the co-authorship of Bose et al. (2017) by members of the quantum collapse community), so appears hardly worth refutation! (Similarly for any view that holds that the gravitational field remains classical.) Equivalently, experimentally ruling out SCG cannot meaningfully increase one's credence that gravity is indeed quantum. We will return to this question in Section 7.

5 Two Paradigms from Fundamental Physics

In this section, we present the two competing "paradigms" which are, in our view, decisive for whether or not one takes the experiments with gravcats to witness the quantum nature of gravity: We will first present the paradigm upon which the experimental setup has to be modeled as bipartite – there is no witnessing of a quantum nature of gravity in this view; in the following subsection, we will then discuss the "tripartite" paradigm, which supports the claim that GIE experiments witness the quantum nature of gravity. For these first two sections, one need not read too much into the word "paradigm"; that these are different approaches one might take to modeling the physics involved in the experiment is sufficient to communicate our point. In Section 5.3, we will argue that they should be understood as paradigms in a Kuhnian sense, as in some way rational advocacy of each is "incommensurable" with advocacy of each other.

5.1 The Newtonian Model

We have already seen that, on a naive account, the Hamiltonian (4.7) will produce entanglement, without assigning quantum states to the gravitational field. As we noted, such a view treats gravity as a direct, instantaneous interaction at a distance; if one takes seriously finite relativistic propagation, and indeed the dynamical nature of the gravitational field in GR, then such an interaction cannot be allowed strictly speaking. But one might come at things from a rather different starting point, and conclude that the naive account is vindicated. Here, the thought is that the treatment of Newtonian gravity as a direct, instantaneous action-at-a-distance interaction, as in the naive model, is merely an apt *stand-in* for the "true" (more on this momentarily) physical degrees of freedom, given that the experiment is assumed to take place entirely in the Newtonian regime. Specifically, Anastopoulos and Hu (2014, Section 3.1) start with the GR action for a scalar field

$$S[g, \phi] = \frac{1}{8\pi G} \int dx^4 \sqrt{-g} \left(R - \frac{1}{2} (\nabla \phi)^2 - \frac{1}{2} m^2 \phi^2 \right), \tag{5.1}$$

where R is the Ricci scalar, and ∇ the covariant derivative. They take a 3+1 Minkowski background, assume linear perturbations, and work in the Hamiltonian framework. They make a standard gauge choice – the longitudinal component of the metric perturbation and the transverse components of its conjugate momentum both vanish – to obtain the Hamiltonian

$$H = \int d\mathbf{r} \, H_{KG} - G \int d\mathbf{r} \int d\mathbf{r}' \frac{\epsilon(\mathbf{r})\epsilon(\mathbf{r}')}{|\mathbf{r} - \mathbf{r}'|} + \dots \tag{5.2}$$

where H_{KG} is the field action in Minkowski spacetime, and $\epsilon(\mathbf{r})$ is the energy density of the field. In the static situation of the GIE experiment, the matter field has no kinetic energy, and the mass-energy contributes only an irrelevant overall phase, so the first term is ignored. The second term yields the Newtonian potential once the nonrelativistic limit is taken, but at this stage of the analysis can be seen to represent, in the chosen gauge, the first-class "scalar" constraint of GR. Finally, the higher-order terms indicated by ellipses include the contribution to the Hamiltonian from the transverse-traceless components of the gravitational field, the so-called "true" degrees of freedom in the chosen gauge; quantizing these yields gravitational quanta, gravitons in this approach. However, to the relevant order, static gravcats produce no perturbations and such terms can also be ignored in the derivation of the interference effect (though not absent from a full description of the experiment, as we discuss shortly). In short, on quantization, the whole Hamiltonian reduces to (4.7).

Following Anastopoulos and Hu (2014), we extract two points from this discussion. The first is simply that this derivation justifies treating gravity as an immediate interaction between gravcats in the proposed experiments: The Newtonian model is simply what is obtained from standard quantization procedures applied to GR with scalar fields, which is itself a strategy well motivated by the success of GR and QFT. The more interesting second point, however, is that the derivation provides us with an analysis of the nature of the degrees of freedom at play: In particular, in the derivation of (5.2), all "true degrees of freedom" of the gravitational field contribute to terms dropped as negligible. By contrast, the only significant term in the Newtonian model is the Newtonian potential, which is not associated to any dynamically propagating parts of the field: The Hamiltonian is nothing but the scalar gauge constraint, and so is "pure gauge." Anastopoulos et al. (2021, Section 6.1) put the idea quite starkly: "Our analysis shows that, *according to GR*, the two parts of a quantum bipartite system that interact gravitationally in the Newtonian regime *do so without an intermediate degree of freedom*" (emphasis in original).

The close analogy (e.g. Chen et al. [2022]) with the more familiar case of quantum electrodynamics may be helpful to reinforce the picture. One way to quantize the electromagnetic field is to perturb around the Coulomb field of the charges, representing the Gauss constraint (the perturbations describing the transverse parts of the potential). On quantization then, the perturbations become photon excitations above a vacuum state represented by the Coulomb potential (e.g. Kleinert [2016, Section 12.3]). In the electromagnetic analog of the GIE experiments, the whole effect comes from the Coulomb – pure gauge – part, the zero excitation state of the true degree of freedom, the photon. Again, it seems that the interestingly quantum part is irrelevant. (Indeed, because of the close formal correspondence, it makes sense to speak of the scalar constraint as a gravitational Gauss constraint. And so we will.)

However, one must be cautious not to lean too heavily on the (widely accepted) technical distinctions between "true degrees of freedom" and (mere) "pure gauge degrees of freedom." Physical significance, even in gauge theories, need not be reserved just for the former (for which one may derive equations of motion). That is to say, "true" degree of freedom, in contrast to "pure gauge," is a standard, well-defined formal notion in gauge theory. Nonetheless, the locution "true" is dangerous: In the quoted remark previously discussed, Anastopoulos et al. appear to move from "true" degrees of freedom in the technical sense to an ontologically thick use of "degrees of freedom" in describing the relevant physics. This move is too hasty. Granted, in linear gravity, the Newtonian potential is pure gauge; but to infer from that formal fact that it is somehow unreal or unphysical is to say that Newtonian gravity – which after all was our best theory of the gravitational field before relativity – itself is therefore "unreal" or "unphysical." Such a view seems implausible[26] (though one could envision it having defenders). Thus, we reject the inference from not "true" in the technical sense to "unreal" in the ontological sense; although we will see later that the split has some relevance to the debate. To avoid any possible conflation of a technical term with an ontological concept, we will thus simply avoid talk of "(true) degrees of freedom" in the following, except on a couple of occasions in which we specifically refer back to this discussion. Indeed, in the paradigm to which we now turn, one simply does not proceed in analyzing the experiments by means of the interpretational tools used in gauge theories.

---

[26] In particular, on quantization, constraints become operators, as we have seen, and so represent physical reality. Christodoulou and Rovelli (2019) offer other arguments that pure gauge interactions are physical. Moreover, as Chen et al. (2022) note, in general, the split between the constraint and dynamical parts of the field is both frame and gauge dependent.

Now, as Chen et al. (2022) demonstrate, it is not correct that the static Newtonian potential suffices for a full description of a GIE experiment. Our treatment focuses on the stage of the experiment in which superposed gravcats interact gravitationally while remaining (approximately) stationary (Marletto and Vedral [2019, Question 5] make the same point). However (footnote 21 for more), before that stage they must evolve into superposition, and after that stage they must evolve out of superposition. During those phases, because the packets are moving, the static field no longer suffices to describe the gravitational interaction, and graviton modes are involved. This is true even though the motion is slow (as it must be, to avoid emitting gravitons, which would destroy gravcat entanglement, as discussed in Section 6).

While we are sympathetic to this point, it does not establish that the Newtonian model is ill-suited (or even that the GIE experiment indisputably witnesses QG, as the tripartite analysis to be discussed shortly has it). For no one doubts that any analysis involves some idealization of the situation. Certainly this is true of the interaction stage: The proponent of the Newtonian model paradigm could take the attitude that we should idealize the previous and following stages as involving no interaction at all. The relevant issue is rather which idealizations to make – a matter of theoretical perspective, which in turn is precisely what is under contention.[27] Of course, if the experiments become sufficiently sensitive that one cannot idealize in this way, then the issue will become empirical. But in that case, one would be in the regime in which one is witnessing gravitons, precisely the benchmark for positive claims of witness that we understand proponents of the Newtonian model paradigm advocate. (Moreover, by measuring that effect one would have moved beyond the GIE experiment.)

---

[27] In correspondence, Charis Anastopoulos proposed time-dependent Newtonian gravity potentials as another available idealization, familiar from work in celestial mechanics, as a reply to the charge that the Newtonian description of the experiment only really applies while the setup remains static. Of course, one requires additional physical argument about the form of the time-dependent potential through the whole duration of the experiment (including outside of the static period), but this is just our point. Different analyses of the underlying physics will, given the specifics of the experimental setup, motivate the different choices of idealization.

5.2 Tripartite Models

Notwithstanding the derivation of the nonlocal interaction term in the Newtonian model in a gauge theoretic approach to GR, there is another view that insists on gravity, in the light of GR, as *mediating* interactions between massive systems such as the gravcats: Specifically, the Hamiltonian should contain no interaction term directly between gravcats, but only interactions between the gravcats and a third "gravity" subsystem. Sometimes this assumption is called "locality"; it can be secured by insisting that gravitational interactions propagate at a finite velocity. However, because considerations of causal connectibility will be relevant in the next subsection, and because the assumption could hold even if the gravcats were spatially coincident (or if effects propagated instantaneously), we will avoid that terminology. What matters here is that the *only* interaction terms are between gravcat and gravity subsystems, and *no* gravcat-gravcat interaction terms.

Under these assumptions, the state of the system after splitting is in fact not (4.1), but

$$|L\rangle \otimes |\gamma_{LL}\rangle \otimes |L\rangle + |L\rangle \otimes |\gamma_{LR}\rangle \otimes |R\rangle + |R\rangle \otimes |\gamma_{RL}\rangle \otimes |L\rangle$$
$$+ |R\rangle \otimes |\gamma_{RR}\rangle \otimes |R\rangle. \tag{5.3}$$

$|\gamma_{XY}\rangle$ represents the gravity subsystem for a pair of gravcats located at X and Y, respectively, so each gravitational state is that appropriate to the gravcat positions in the corresponding term. After t, the state is not approximated by (4.3) but by

$$|L\rangle \otimes |\gamma_{LL}\rangle \otimes |L\rangle + |L\rangle \otimes |\gamma_{LR}\rangle \otimes |R\rangle + e^{\frac{-iGm^2t}{\delta}} |R\rangle \otimes |\gamma_{RL}\rangle \otimes |L\rangle$$
$$+ |R\rangle \otimes |\gamma_{RR}\rangle \otimes |R\rangle.^{28} \tag{5.4}$$

The key thing to recognize is that in states (5.3) and (5.4), the gravity subsystem is in a *superposition* of classical gravitational states. This is of course because the gravcats are in a spatial superposition and the gravity

---

[28] And (see footnote 21) at the end of the process, after passing back through the Stern–Gerlach device, the full state can be written

$$\left(|X\uparrow\rangle \otimes |Y\uparrow\rangle + |X\uparrow\rangle \otimes |Y\downarrow\rangle + e^{\frac{-iGm^2t}{\delta}} |X\downarrow\rangle \otimes |Y\uparrow\rangle + |X\downarrow\rangle \otimes |Y\downarrow\rangle \right) \otimes |\gamma_{XY}\rangle.$$

subsystem couples to the wavepackets through the Hamiltonian.[29] Thus, in this model, the gravity subsystem exhibits a quantum nature, in the sense of superposing classical states.

Exactly what the mediating quantum gravitational system is taken to be is not the same across theoretical treatments: For Marletto and Vedral (2017a), it is an information channel; for Bose et al. (2017) and Chen et al. (2022), a quantum field; and for Christodoulou and Rovelli (2019), a metric field capable of quantum superpositions, and so on. Further, Adlam (2022), referring to our discussion of tripartite models, distinguishes "thin" and "thick" variants: the former capturing minimal aspects of the nature of gravity, and the latter including central features of GR. Of course, ultimately, they all take these to be different representations of the same physical object.

We consider some of these ideas in more detail in Section 5.2.2, but first we will see that the production of gravcat entanglement witnesses the quantum nature of gravity (in tripartite models) on quite general grounds. *All* that needs to be assumed in fact is that it is a mediating subsystem, in line with the general conception of gravity in GR. We already saw in Section 4.3 that if one takes view 2 of SCG – SCG as a candidate fundamental theory – then we should expect no entanglement in the experiments with gravcats. However, we will now see that this result can be generalized to the conclusion that no classical mediator can produce entanglement between quantum systems with no direct interaction.

5.2.1 Quantum and Information Theoretic Considerations

In this section, we present two results, showing from general quantum mechanical and general information theoretic considerations, respectively, that interactions with a classical mediator cannot lead to entanglement between two systems. Two preliminary notes. First, of course, the very setup of two local subsystems and a mediating subsystem means that the tripartite paradigm is assumed from the beginning, so these results

---

[29] To keep things general, we will not specify a Hamiltonian, but given our assumptions, it must have the following effect: There is a continuous evolution, not only of the gravcats, but also of the gravity subsystem. Moreover, we assume in this experiment that time scales are long compared to D/c.

cannot settle the tripartite-Newtonian question. What they can do is show why the observation of entanglement in a GIE experiment means that the tripartite system must have a quantum mediator – or rather, as we shall see, that the mediator not be classical. Second, results like those we discuss presently are given for simple, finite-dimensional systems. One might question the generality of such analyses in the context of infinite-dimensional gauge field theories – a question we set aside.

First then, a proof in quantum theoretic terms.[30] We will assume the simplest kind of system, in which the subsystems are two-dimensional, but as Marletto and Vedral (2017a) discuss, the extension to N-dimensional systems seems straightforward, and thence to quantum fields and the GIE experiment in the Fock representation (although this claim has been challenged: Anastopoulos and Hu [2022, Section 2]).

> **Claim:** Suppose that $\mathcal{H}_2$ is a two-dimensional Hilbert space, and consider the tripartite system $\mathcal{H}_2 \otimes \mathcal{H}_2 \otimes \mathcal{H}_2$. (a) Let the first and last slots describe qubits that interact "locally": the Hamiltonian $H = H_{12} \otimes I + I \otimes H_{23}$ (where H_{12} acts on the first and second terms of the tensor product only, etc.). And (b) let the second slot represent a classical bit: The only observable for this subsystem is σ_z (and the identity I). Given the locality of the qubits as expressed by (a) and the classicality of the bit as expressed by (b), no entanglement can arise between the qubits.

Proof:

1. From (a) and (b) we have

$$H_{12} = A \otimes I + B \otimes \sigma_z \tag{5.5}$$

$$H_{23} = I \otimes C + \sigma_z \otimes D \tag{5.6}$$

assuming that the Hamiltonian is an observable. It follows that

$$[H_{12} \otimes I, I \otimes H_{23}] = 0 \tag{5.7}$$

As the two terms in the Hamiltonian H commute, the unitary operator describing the time translations factorizes, that is,

---

[30] We thank Richard DeJong, who provided the outline of the following proof, which differs from the original proof by Marletto and Vedral (2017a) (though was inspired by it, and by a conversation with Marletto). Many of the papers in the literature simply point to the theory of "local operations and classical communication" (LOCC); see Christodoulou et al. (2022) for a clear statement.

$$e^{-iHt} = e^{i(H_{12} \otimes I + I \otimes H_{23})t} = e^{-iH_{12} \otimes It} \cdot e^{-I \otimes H_{23}t}. \qquad (5.8)$$

That is, since the two factors commute, we can treat the evolution as an interaction first between one of the qubits and the bit, and then between the other qubit and the bit, in either order.

2. So let the tripartite system start in a fully factorizable pure state, Ψ, and apply the unitary evolution:

$$e^{-iHt}|\Psi\rangle = e^{-iH_{12} \otimes It} \cdot e^{-I \otimes H_{23}t}|\psi_1\rangle \otimes |\psi_2\rangle \otimes |\psi_3\rangle$$

$$= e^{-iH_{12} \otimes It}|\psi_1\rangle \otimes \sum_{j=0}^{1} \alpha_j |\chi_j\rangle \otimes |\phi_j\rangle, \qquad (5.9)$$

where in the last step, we have used the fact that the bit is classical (b), so there is only one orthonormal basis for the mediating subsystem, namely $\{|0\rangle, |1\rangle\}$, respectively, the -1 and $+1$ eigenstates of σ_z. Hence we can continue

$$= e^{-iH_{12} \otimes It}|\psi_1\rangle \otimes \sum_{j=0}^{1} \alpha_j |j\rangle \otimes |\phi_j\rangle. \qquad (5.10)$$

3. Since (b) the only observables for the bit are (multiples of) the identity and σ_z, there is a superselection rule: *All* observables on the system commute with $I \otimes \sigma_z \otimes I$. In particular, restricting attention to the bipartite 2–3 subsystem, the most general form of an observable is $I \otimes A + \sigma_z \otimes B$, which commutes with $\sigma_z \otimes I$. As a result (which can also be readily checked), the pure state $\sum_{j=0}^{1} \alpha_j |j\rangle \otimes |\phi_j\rangle$ is indistinguishable from the mixture $\sum_{j=0}^{1} |\alpha_j|^2 |j\rangle\langle j| \otimes |\phi_j\rangle\langle\phi_j|$: The expectation values of the states agree for all observables of the general form.

Therefore, we can replace the pure state in (5.10) with a physically equivalent density matrix,

$$|\psi_1\rangle \otimes \sum_{j=0}^{1} \alpha_j |j\rangle \otimes |\phi_j\rangle \rightarrow |\psi_1\rangle\langle\psi_1| \otimes \sum_{j=0}^{1} |\alpha_j|^2 |j\rangle\langle j| \otimes |\phi_j\rangle\langle\phi_j|.$$

$$(5.11)$$

This state is separable between 2 and 3 (and indeed between 1): In general, any density matrix of the form $\sum_j |c_j|^2 \rho_j^a \otimes \rho_j^b$ can be expressed as a mixture of factorizable pure states. In short, by rewriting the state in this way, we have made manifest the absence of entanglement between 2 and 3, ultimately *a consequence of the classical nature of the bit*.

4. Now, it should be clear that since 3 and 2 are not entangled, then – even if 2 were quantum – an interaction between 1 and 2 alone cannot entangle 1 and 3. (Of course, if 2 and 3 were entangled, that entanglement could be "exchanged" between 2 and 1.) But for completeness, we will show this to be the case. So consider the interaction between 1 and 2, producing the state

$$e^{-iH_{12} \otimes It} |\psi_1\rangle\langle\psi_1| \otimes \sum_{j=0}^{1} |\alpha_j|^2 |j\rangle\langle j| \otimes |\phi_j\rangle\langle\phi_j| e^{+iH_{12} \otimes It}$$

$$= \sum_{j=0}^{1} |\alpha_j|^2 e^{-iH_{12}t} |\psi_1\rangle\langle\psi_1| \otimes |j\rangle\langle j| e^{+iH_{12}t} \otimes |\phi_j\rangle\langle\phi_j|. \quad (5.12)$$

This state is a mixture of pure states in which the 1–2 subsystem factorizes with the 3 subsystem; hence it is separable between the two qubits. (Of course, by appeal to the classicality of the bit, we can as before show that it is also separable between the 1 and 2 subsystems.) Thus, we have shown using the standard assumptions of QM and the locality of the qubits (a), that no entanglement can arise between the qubits, given the classicality of the bit as expressed by (b). □

Now, as we noted in Step 3, there is a superselection rule, so that the bit state is always a mixture of $\sigma_z = \pm 1$ states. But (since we assume the Hamiltonian to be an observable), the superselection rule implies a selection rule. That is,

$$[I \otimes \sigma_z \otimes I, H] = 0 \quad (5.13)$$

so that the value of the bit is a constant. Thus not only is the bit a mixture of +1 and −1 σ_z states, it is a *constant* mixture. In other words, in this framework, the qubits have no effect on the bit, although the bit state does affect the final states of the qubits. The one-way nature of the interaction seems to refute the claim that the bit mediates any interaction *between* the qubits, after all – so it is no surprise that no entanglement arises between them! (Of course, if entanglement *does* occur between local systems in the sense of (a), as expected in the GIE experiment, then the proof shows that (b) fails, so the mediator is nonclassical. In particular, there are then no selection or superselection rules, and so the mediator is not frozen in some mixed state.)

This situation makes one wonder whether the quantum framework is actually apt for representing a classical bit in the first place. Perhaps the conclusion that the mediator cannot be classical depends on adopting a hobbled description for it. One would like to see the same result in a broader setting that doesn't already suppose the quantum formalism. Moreover, the proof assumes in Step 1 that the Hamiltonian is an observable, and one might somehow question that assumption. For these and other reasons, it is worth sketching an alternative, more general information theoretic result.

Marletto and Vedral (2020) use the general information theoretic framework of "constructor theory" (Deutsch and Marletto 2015) to prove that if interactions with a mediating subsystem produce entanglement between mutually isolated systems, then the mediator cannot be classical. Both the exact statement of the theorem and its proof rely on some new and unfamiliar machinery, so we will just sketch some central ideas to give a feeling for the result; for details, the reader is referred to the original papers.[31]

Constructor theory works at a very high level of generality, in which various subsystems are posited, each with some set of mutually exclusive states s_i, thought of as providing a full description of the subsystem. In addition, state transitions for individual and composite systems – "tasks" – $[s_i \rightarrow s_j]$ are specified as possible or impossible. For instance, suppose we have a subsystem S with states $\{0, 1, y, z\}$ (the numerals and letters should be understood here simply as state labels). Sets of states correspond to determinables (or "variables" in constructor theory) and individual states to their determinate values ("attributes").[32] Let us suppose then that $B = \{0, 1\}$ is one variable, with attributes 0 and 1 (so that B is for 'bit'); and that $X = \{y, z\}$ is another.

---

[31] A theorem with analogous conclusions has also been given in the framework of generalized probability theory (Galley et al. [2022]). We focus on the constructor theory result for definiteness, but stress that the availability of similar results in other frameworks indicates the generality of the underlying argument.

[32] Since, in general, one wants to allow for degenerate states, properly speaking attributes are sets of states, and variables are sets of disjoint attributes. In our examples, we will simplify the notation by ignoring this extra structure.

Simplifying considerably, let us say that a variable $\{x_1, x_2, \ldots, x_n\}$ of a system is an "observable" when there is a second system – the measuring apparatus – with states $\{s_r, s_1, \ldots, s_n\}$, and a possible task that for any $1 \leq i \leq n$ has the effect $[(x_i, s_r) \rightarrow (x_i, s_i)]$. That is, s_r is the apparatus-ready state, the s_i are its various possible meter readings, correlated with states of the subsystem, so that the task amounts to a nondestructive measurement of the variable (the state of the measured subsystem is unchanged).

To continue our example, let us thus further suppose that B and X are both observables. In that case, we have a further choice: Their union may or may not also be an observable – in describing a system, we choose whether or not the necessary process is possible. If $\{0, 1, y, z\}$ is an observable, then S is simply a classical system, with a four-dimensional state space. But if it is not an observable, then the situation is like that in a qubit, in which B and X are incompatible observables, σ_z and σ_x say, which cannot be observed simultaneously. Broadly speaking then, this situation abstracts the notion of "complementarity" without assuming the quantum formalism. So we see that constructor theory can accommodate classical and quantum on an equal footing; the general framework of states and tasks is neutral between them, and the difference arises only from which tasks are possible. (Of course, quantum is also distinguished from classical by the capacity for entanglement, beyond any classical statistical correlation, and this distinction too can be represented.) Moreover, the generality of the framework allows for possibilities between the fully classical and the fully quantum (and indeed possibilities orthogonal to that distinction); in particular, the generalizations of complementarity and entanglement, while clearly nonclassical, do not by themselves entail the full quantum formalism.

Suppose, then, two mutually isolated qubits (or "superinformation media", as they are represented in constructor theory), which can each interact with a mediating subsystem: Of course, this state of affairs is cashed out in terms of which two subsystem tasks are and are not possible. Then, Marletto and Vedral show, if initially separable qubits end up entangled, then the mediator must be nonclassical, possessing complementary variables, and entangling with the qubits, in their generalized senses.[33] Thus, given the tripartite model, we again conclude that a positive result in the GIE experiment would witness nonclassical behavior of gravity. In this framework, we don't shoehorn a classical system into QM, and there is no requirement that the mediator be static if classical.

Note that strictly the theorem shows that mediating entanglement requires a nonclassical mediator, not that it requires a quantum one: indeed nonclassical, nonquantum models accounting for entanglement mediation exist (Marletto and Vedral (2019) refers to Hall and Reginatto [2018]). In this regard, Marletto and Vedral reasonably compare their result with Bell's theorem: In that case too, experiment can only show that a system is nonclassical, not that it is quantum.

Now that we have seen how entanglement mediation rules out classicality, we should note that GIE experiments also bear on gravitational collapse theories (often associated with Diósi (1989) and Penrose [1994]). Such proposals could be thought of as falling into the tripartite camp, but claiming that the gravitational field abhors a superposition, and induces a collapse whenever it deviates from a classical state (by a prescribed amount), and thus is effectively always classical. Here, the implication of observing entanglement is more equivocal than for SCG: As Bose et al. (2017, p. 1) put it, such theories imply "the breakdown of quantum mechanics itself at scales macroscopic enough to produce prominent gravitational effects." The question of course is what counts as "prominent." On the one hand, by Penrose's estimates, the proposed experiment, with gravcats of 10^{-14} kg separated by $100 \, \mu$m, the gravitational collapse time should be of the order of a second, which would be fast enough for the classicality of the field to affect any observed entanglement. And so it seems it is a "prominent" effect: The quantum state will collapse, and no entanglement will be observed. However, on the other hand, should entanglement be observed, the theories do have a tunable parameter, which could be set to prevent collapse in the currently envisioned GIE experiments, although they would place a

---

[33] We note that the theorem requires the union of the complementary variables not only to be unobservable in the sense we gave, but to not even be measurable by a destructive measurement that changes the state of the mediator. Nor is it required that both the variables be observables; one of them could similarly be "hidden"; in other words, since it is complementary to the other variable, it could be a nonclassical hidden variable.

new bound on it. But so doing is to accept that the experiment witnesses a quantum superposition of the gravitational field, which is at least against the spirit of Penrose's position, and quite possibly falls afoul of the very arguments by which he motivates it.

In this context, we also observe that the GIE experiments are of a different character from that envisioned in Marshall et al. (2003) to explore (inter alia) collapse theories, including the Penrose–Diósi type. At the heart of the setup is a single gravcat in the form of a mirror: It is put into spatial superposition by reflecting the two packets of a beam split photon off it; and one observes whether it remains in that state, or collapses because of its sizable gravitating mass. If collapse were observed, that would be evidence that GIE would not be observed, because the gravcats in the GIE experiments would not remain in superposition either. But in the alternative case, it would not require any quantum behavior of gravity to explain an observed absence of collapse; that outcome would be compatible with SCG even. In this regard, then, the experiment asks a very different question to the GIE experiments.

Finally, given this relevance of GIE experiments to collapse interpretations, and indeed to SCG in a more interpretation-neutral way than Page and Geilker, it is interesting to note the claims of Adlam (2022) that GIE experiments can function as *evidence for* interpretations of QM, given the tripartite paradigm. Adlam starts with a standard distinction between a ψ-complete interpretation of matter theories (which only includes a quantum sector), and a ψ-incomplete interpretation (which also includes "non-quantum be-ables"). ψ-incomplete interpretations come in further variants: Either the matter theory's ontology includes the quantum sector but also some other nonquantum sector (ψ-supplemented); or its ontology includes only a nonquantum sector, so the quantum sector is more than epistemic but not ontological (say lawlike structure); or purely epistemic (ψ-epistemic). Her main argument is then that, given that on ψ-incomplete views gravity could also be sourced by a nonquantum sector (and, in the case of ψ-epistemic interpretations, may even have to be thus sourced), there seems a correlation between a negative result on spacetime superpositions in the context of tabletop experiments, and increased confidence in a ψ-incomplete interpretation. Conversely, a positive result on spacetime superpositions from the tabletop experiments would seem to boost confidence in a ψ-complete interpretation.

5.2.2 The Quantum Description of Gravity

In this subsection, we elaborate further on the quantum description of the "gravity" subsystem in these experiments, within the tripartite models paradigm, which explicitly views gravity as a physical subsystem on a par with the gravcats. The questions are, what, formally and materially, should the states $|\gamma\rangle$ be, what are the gravcat states, and what is their interaction? It is of course natural to start with GR: Our current best physical theory of gravity after all does theorize gravity as a dynamical system with a nontrivial vacuum sector. So, within the tripartite models paradigm, GR itself would seem to justify the assumption in (5.3) that we have a tripartite system, something that seems ad hoc in Newtonian gravity proper. And indeed, this is the starting point that theorists have generally taken, in a few different ways.

For instance, the supplementary materials to Bose et al. (2017) derive the phases of the various terms in the gravcat superposition by quantizing linearized GR: Briefly, one again starts by representing the classical GR gravitational field as linear perturbations in a 3+1 Minkowski background, using the stationary approximation for the gravcats, but because the gravcats are nonrelativistic, only the energy-momentum, T_{00}, component of the stress energy contributes. So only the h_{00} components of the field are relevant, and hence quantized. The experiment can then be modeled in terms of an interaction between localized mass excitations, mediated by coherent states of the modes of the h_{00} field.[34] The appropriate Hamiltonian has been solved, and introduces the same phases as predicted by the Newtonian potential.

Another approach is taken by Christodoulou and Rovelli (2019), who model the experiment using quantum superpositions of the metric field, locating the experimental effect as a consequence of superpositions of gravitational redshifts. Of course, from a quantum field theory perspective, these are presumably superpositions of coherent states of the perturbatively quantized (linearized) gravitational field, but in their presentation they are simply treated as quantum states labeled by classical

---

[34] As discussed in Huggett and Wüthrich (2020, chapter 9), coherent states are often taken to represent classical states of quantum fields. Note that what are usually called gravitons are associated with other components of the h field.

geometries.[35] Starting classically, again assuming that the gravcats are heavy enough, and the duration short enough that they remain (approximately) stationary during the experiment, the metric field due to a body of mass m and radius R can be approximated by

$$ds^2 = (1 - 2\phi(\vec{r}))dt^2 - d\vec{r}^2. \tag{5.14}$$

Outside the body, $\phi(\vec{r}) = Gm/r$ is just the Newtonian potential; while inside the body, one takes $\phi(\vec{r}) = Gm/R$, a constant. (Compared with (3.1), we now set $c = 1$ and consider the interior as well as exterior field.) As always, we are in the regime of linearized gravity, so that for a pair of gravcats, the classical geometry will be the sum of two such solutions. It is then straightforward to calculate the proper time elapsed along the worldline of (a point in) one gravcat at a distance $\delta \gg R$ from another:

$$\tau = \int_0^T ds = \int_0^T dt\sqrt{1 - 2Gm/R - Gm/\delta} \approx T \cdot (1 - Gm/R - Gm/\delta),$$
$$\tag{5.15}$$

showing the usual GR time dilation effect of a nearby mass. If $d \gg R$, most of the effect will come from the mass of the gravcat itself, but we also see that there is a contribution dependent on the distance to the other gravcat.

Thus, the components of the superposition (4.1), corresponding to different gravcat separations, correspond to different metrics, and hence different proper times. Now assuming that the (classical) gravitational

---

[35] As will become apparent, the approach to be sketched is somewhat heuristic. Chen et al. (2022) give a more rigorous treatment in the h_{ij} field basis of quantized linear 3+1 gravity; essentially, the gravitational states then are a quantum superposition of the classical linear gravitational field (that of the Fierz–Pauli action) in a Minkowski background. This treatment also resolves any worries about the tensor product space, due to the gauge nature of gravity. That being said, there is also a more skeptical note of caution in the vicinity: Quantized linear 3+1 gravity might break down in surprising regimes in an underlying QG theory, as perhaps indicated by black hole information considerations (a regime where, save the threat of paradox, quantized linear 3+1 gravity should be safely deployed). It has been suggested in (Raju 2022) that, in the black hole case, what breaks down could be a "split" property in the low-energy description, which underwrites spacetime locality being a guide in that low-energy description for decomposing a system into subsystems. If this is right of the underlying QG theory, one may also fear the split property failing in other surprising regimes like that of the GIE experiments, due to as-yet unknown features of QG.

field labels what is actually some coherent state in a quantum treatment of gravity, which can superpose, the full state is therefore (5.3), with $|\gamma_{LR}\rangle$ and so on describing different states of the metric field of GR. And thus the different terms of the superposition are associated with different proper times and will develop relative phases (relative to a common lab time, T). In particular, a gravcat pair has a phase $e^{-im\tau}$ from its mass energy, or (ignoring a common phase) $e^{-iGm^2T/\delta}$ from (5.15) – the very phase one expects (4.3) from the Newtonian potential, but now seen to be a relativistic redshift effect. Thus, the very same relative phase between the components of the gravcat superposition, and ultimately the very same entanglement, is predicted by treating metric states as superposable quantum states (to leading order).

Both of these treatments, which in different ways "quantize" the metric field of GR, relate gravity to a physical quantity (a metric field!), which naturally should be treated as a third subsystem in addition to the gravcats, and which then must be quantum (or at least nonclassical) by the general quantum and information theoretic arguments, if the gravcats become entangled as expected. Thus, from this point of view, the derivation of the effect by appeal to the Newtonian potential with which we presented it on the naive first pass is an approximate heuristic. (Recall for contrast that the analysis by Anastopoulos and Hu in chapter 5 – or rather the interpretation thereof in favor of the Newtonian model view – explicitly renders the Newtonian potential as uninformative of the physical parts of the gravitational field.)

A word of caution: Because modeling the system in this way involved an appeal to GR, it is tempting to think that the finite speed of propagation of the gravitational interaction in GR is necessary and sufficient for the system to be tripartite. But while a finite propagation speed seems to require that the system be tripartite, we would nonetheless now like to stress that it is by no means necessary. This point can be seen clearly by adopting a geometrized formulation of Newtonian gravity: Newton–Cartan theory.[36] In fact, as we demonstrate in the appendix, whereas the gravitational redshift effect noted in Christodoulou and Rovelli (2019) is

---

[36] There is a robust tradition in the foundations of physics of moving to the geometrized Newton–Cartan theory, in order to more sharply compare the features of Newtonian gravitation with the features specific to its relativistic successor theory: for example,

specific to their relativistic setting, the surrounding argument – namely, that a superposition of spacetime geometries demonstrates the quantum nature of gravity – is not. In particular, one can perform an independently motivated analysis of the experiments with gravcats in the framework of Newton–Cartan theory, which is analogous to their relativistic analysis: In both, gravcat pairs develop different phase factors with respect to different, superposed, spacetime geometries, leading to entanglement. Again, all that is important is the decision to treat gravity as a mediating subsystem, a decision whose subsequent execution proceeds analogously across both theoretical contexts.[37]

Moreover, this example emphasizes that the mediating subsystem need not be a dynamically propagating field for the entanglement theorems to apply, and hence for the GIE experiment to witness QG.[38] In other words, even though the theorems entail that a successful GIE experiment witnesses the nonclassical nature of gravity, it does not, without further assumptions, witness perturbatively quantized linearized gravity. Of course, it is possible that these further assumptions are supplied by the concurrent commitment to GR in the background of an analysis of the Newtonian-regime experiment. This is a point we will discuss in Section 6, in the context of a thought experiment that has been analyzed by Belenchia et al. (2018).

Finally, it is worth pointing out that the locality assumption behind the tripartite view is only to hold approximately at the relevant energy level of the experiment – underlying physics (say, at a string scale) may just as well be nonlocal. More precisely, so it has been explicitly shown by Marshman et al. (2020, Section VII) for modifications of gravity in the UV based on the most general quadratic actions in four dimensions that are invariant under parity and torsion-free – such actions generally

---

Misner et al. (1973, chapter 12), Malament (1995, 2012, chapter 4), Weatherall (2014), Ehlers (2019).

[37] We note that those central features of GR that constitute the thick variant of the tripartite view according to Adlam (2022) are also features of this Newton–Cartan treatment: namely that the gravitational subsystem may be described geometrically.

[38] For a quantum treatment of Newton–Cartan gravity, see Christian (1997). From another point of view, as we will discuss later, Christodoulou et al. (2022) offer a description of the GIE experiments as a degenerate case of a local, retarded relativistic model where retardation/causality is neglected, so that the mediator is again represented as nondynamical.

simply include nonlocal interactions. In a sense, this point is not very surprising: Even beyond the effective field theory paradigm, we tend to hold physics at some domain to be independent (i.e., decoupled) from that at lower scales (higher energies). Interestingly, for future experimentation, however, a disagreement in underlying physics (including an actual nonlocal nature of gravity) becomes eventually distinguishable through the relevant entanglement entropy in the experiments with gravcats when performed with ever smaller probe separations.

5.3 Why "Paradigms"?

At this point, we should clarify and justify our description of the Newtonian model and tripartite models as, respectively, not just views of the GIE experiments, but *paradigms*. First, by this term, we do not mean to commit to a bundle of all the senses and aspects of "paradigm" inaugurated in Kuhn (1962); and certainly we do *not* claim that there is incommensurability between the paradigms, understood as an incommensurability in either meaning or standards of experiment, let alone in the very nature of rationality. Rather, we have in mind a sense of paradigm developed by the Kuhn of *Objectivity, Value Judgement, and Theory Choice* (in Kuhn [1977a]), who emphasizes that scientists can agree about the criteria of rational choice, yet come to different decisions because they disagree about how to implement them. For instance:

> When scientists must choose between competing theories, two men [sic] fully committed to the same list of criteria for choice may nevertheless reach different conclusions. Perhaps they interpret simplicity differently or have different convictions about the range of fields within which the consistency criterion must be met. Or perhaps they agree about these matters but differ about the relative weights to be accorded to these or to other criteria when several are deployed together. With respect to divergences of this sort, no set of choice criteria yet proposed is of any use. One can explain, as the historian characteristically does, why particular men made particular choices at particular times. But for that purpose one must go beyond the list of shared criteria to characteristics of the individuals who make the choice. One must, that is, deal with characteristics which vary from one scientist to another without thereby in the least jeopardizing their adherence to the canons that make science scientific. Though such

canons do exist and should be discoverable ..., they are not by themselves sufficient to determine the decisions of individual scientists.
For that purpose the shared canons must be fleshed out in ways that
differ from one individual to one another. (324–325)

Here, Kuhn argues that no calculus of theory choice will universally
determine the unique correct theory to adopt in the light of evidence
and extra-empirical criteria. The canons of scientific rationality, that is,
leave room for reasonable people not only to disagree, but to recognize
one another's rationality in reaching different conclusions. (Of course, in
actuality, they might not, either because they do not appreciate Kuhn's
point, or because they abandon the scientific canons.)

However, Kuhn's specific analysis is not a perfect fit for the present
case, because in the GIE experiments we are not faced with disagreement over basic theory, but over how that theory should be used to model
a particular proposed experimental phenomenon. That is, as we have
emphasized, both models take as their starting points the methodolog-
ical principles that a model of GIE should start from our best theory
of gravity, GR, and the application of generally accepted methods of
quantization, namely quantized linear gravity, understood as an EFT.
Such assumptions are very widely (though not universally) held in the
QG theoretical community, so the different models do not represent a
different understanding of what is "really" going on, at a more fundamental level. For instance, the situation is quite different from the choice
between Ptolemaic and Copernican systems in the sixteenth century (to
take Kuhn's example). GIE is even disanalogous to the choice between
Lorentzian and Einstein–Minkowski accounts of the null outcome of
the Michelson–Morley experiment. Even though these cases are alike
insofar as their competing paradigms make exactly the same predictions
so that the choice is strictly underdetermined by the data, they differ
crucially because in the case of GIE, they do not disagree about the basic
underlying physics.[39]

Instead, the GIE paradigms disagree with respect to how to extract
and approximate a model for the experiment, from a shared physics. It
is striking when distinct theories predict exactly the same data, but not

---

[39] Thanks to Karim Thébault for pressing us on this point.

so remarkable that two different models, extracted by established methods from a successful theory, do. Rational indeterminacy arises because each model embodies a different aspect of GR as central: In one case, it is the lesson that gravity is causally propagated, while in the other, it is the way in which one takes the Newtonian limit. Which aspect one attaches the greater weight to, given the particulars of the GIE experiment, determines which model one concludes is the most faithful to the experiment; but one can reasonably take either to be more important, and so shared rational criteria leave the model indeterminate. And while proponents of both models can be brought to *understand* the analysis of the other, they will fail to accept the complete relevance it has for the other – the degrees of freedom analysis of when to call putative gravitational systems physical is simply mutually exclusive from viewing the gravitational field as propagating and vice versa. In this sense, the models are "incommensurable."

It is this commonality with Kuhn's argument that then leads us to conclude that talk of "paradigms" is appropriate here. (At least paradigms "lite," because stronger forms of incommensurability are not in evidence.) The analysis of the models that we have offered in Section 5 has been aimed at making clear the different modeling assumptions behind the models, to reveal the different theoretical stances embodied in the models, and to demonstrate that indeed they are all reasonable, just weighed differently. We hope that doing so, plus our discussion of paradigms, will go some way to helping clarify – though not resolve, obviously! – disagreements over the correct model.[40]

Shortly after this Element appeared on the arXiv preprint repository, so did a parallel analysis (Fragkos et al. [2022]), which identifies the choice of gauge as the "assumption" leading to divergence on the question of Newtonian versus tripartite models. That is, the different choices

---

[40] Samuel Fletcher pointed out one other way in which our use may deviate from Kuhn's. Doesn't LEQG constitute an overarching paradigm, and so a form of normal science, within which the Newtonian-tripartite dispute plays out? If so, however, by definition, *paradigm* disputes only occur when normal science breaks down. It's a fair point, and a useful qualification, but we again point to the incommensurability that we have identified. (Moreover, we note that the point does not apply to someone who adopts the tripartite paradigm, but does not commit to LEQG, as in view 3 of SCG, or an advocate of the LOCC approach who remains neutral on the matter of LEQG.)

of gauge made in Anastopoulos and Hu (2014) and Bose et al. (2022) lead to different models: one in which a Newtonian potential comes from a gauge constraint (as discussed earlier), and a tripartite model in which modes of the field are quantized degrees of freedom (somewhat different from the h_{00} quantization discussed earlier).[41] Technically, this point is correct, and indeed highlights a key decision point; but it is also too narrow, missing the many (other) ways one might give a tripartite model of gravity. Indeed, as Fragkos et al. indicate, a more basic disagreement than any choice of preferred gauge involves a whole package of views one might hold about the nature of Newtonian gravity – views about the importance of causality, constraints, and so on. Our emphasis thus far has been to articulate how these latter views come together as distinct *packages*, and that these packages are the ultimate difference-makers in interpretations of the GIE experiments.

Moving on, suppose that one accepts our application of Kuhn's conception to GIE experiments, one can still worry that we have missed some argument, relying on deeper principles agreed to by the partisans of both paradigms, which would after all rationally require one group to abandon their commitments, leading to the collapse of our talk of paradigms. Our presentations of this work[42] have indeed encountered such resistance to calling the positions "paradigms" in our sense, some rehearsing considerations already discussed, and some raising new arguments to be considered here. It is worth noting, however, that while those arguing against paradigm talk have done so because they believe that one of the models is clearly the right one, they have not agreed about which one in fact is the right one. So this datum is (at least) compatible with the thesis that our interlocutors are adopting different paradigms in our sense! Nonetheless, we take their arguments seriously, and indeed addressing them will help clarify our claim.

Objection: By denying gravity the status of a causal field, the Newtonian model is committed to such an unreasonable refusal of physical

---

[41] They also describe quantizing an absorber-theory reformulation of linear gravity, in the style of the Wheeler–Feynman reformulation of electromagnetism, and explain that it also leads to a no-witness conclusion.

[42] For instance, to audiences at the Center for Philosophy of Science at the University of Pittsburgh, and the Rotman Institute at Western University.

background knowledge that it is disqualified as sensible physical theorizing by any paradigm-independent standard of modern physics.

Reply: The Newtonian model does take into account physical background knowledge – arguably, just the same knowledge, but it weighs the relevance of the analysis given in Section 5.1 completely differently. Indeed, that the Newtonian model takes into account the same physical background knowledge is emphasized, ironically, in a recent article by Christodoulou et al. (2022). In the article, which is intended to reinforce that GIE experiments indicate quantum superposition of spacetimes, the authors distinguish a "slow-motion" approximation (sources moving at nonrelativistic speeds), where the gravitational interaction is still local (i.e. causal), and a "near-field approximation" (sources are much closer together than time elapsed through the experiment, in natural units). The "Newtonian limit" that reproduces the naive model, claim the authors, denotes the overlapping regime where both approximations are satisfied. Now, as they show, in the slow-motion, not near-field approximation regime, locality considerations show up as corrections between the employ of laboratory time function (as in the Newtonian limit) and a retarded time function. But, setting up retarded GIE experiments in the slow-motion, not near-field regime is "not realistic for the foreseeable future" and seeing the effects of slow-motion correction terms to the Newtonian limit analysis of the GIE experiments is "unlikely reachable" (p. 4). Thus, although there are retarded GIE protocols that they point to on a further horizon of experimentation (p. 5) that, given physical background knowledge, would preclude a Newtonian model description, the GIE experiments under scrutiny here do not: precisely because they are in the Newtonian limit regime of our fundamental gravitational physics.[43]

---

[43] Note that the main point of their article is ultimately to develop the physical picture of quantum superposition of spacetimes in GIE experiments generally, which would include the non-retarded GIE experiments in the Newtonian limit: "The physical picture arising from the analysis is that information travels in the quantum superposition of field wavefronts: The mechanism that propagates the quantum information with the speed of light is a quantum superposition of macroscopically distinct dynamical field configurations" (p. 5). On the other hand, Charis Anastopoulos, in correspondence following the initial circulation of this Element, pushed back on exactly this latter conclusion. Drawing on his co-authored (2021) article, Anastopoulos insists that there may be ways to extend the Newtonian paradigm to cover retarded GIE experiments. In this case, the relevant paradigm distinction should rather be between a weak gravity

Objection: Both paradigms take GR as their starting point, thus accepting that gravity is well described by a dynamical classical field in the appropriate limit. What then is the state of this field supposed to be if we have superposed gravcats, and their consequent observable entanglement? The general information theoretic theorems show that it cannot be a classical state if the interaction is mediated by gravity in the appropriate sense, and no other mechanism in which gravity remains in a classical field state has been proposed (if any be possible). Doesn't then entanglement require a quantum superposition according to the tenets of both paradigms?[44]

Reply: First, one might accept GR in its regime, but not accept that gravity is a "field" (broadly speaking), classical or quantum, more fundamentally. One might have a specific alternative theory in mind (gravity as ancilla, perhaps), but more likely, one might simply be neutral on the topic. That would be an unusual stance for a theorist, but would be less implausible for an experimentalist, whose goal of probing the quantum + gravity intersection requires only very broad theoretical commitments (and a far more detailed understanding of the causal powers of the various elements within any specific experiment, e.g. Hacking [1984]). Either way, one could resist the objection, because one is not committed to any field state at all.

However, as we noted, most theorists accept the EFT approach to QG (including of course anyone accepting the Newtonian analysis of Anastopoulos and Hu [2014]), and thus that "really" gravity is a field, and so a third subsystem as conceived in the tripartite approach. Isn't any such a person thus bound to accept that paradigm, and conclude that gravcat entanglement would witness a nonclassical state of the field? This thought leads to the second reply: "yes and no." The "really" provides important wiggle room, because it means "(more) fundamentally"; in which case yes, in a more complete physical description, gravity is in a superposition, but also perhaps no, in the model *most apt* for the experiment, gravity is not represented as quantum, and hence such behavior

----

regime including relativistic speeds (and not merely a Newtonian sector) and the tripartite models view, the reason being that the weak gravity regime including relativistic speeds is also adequately captured by gauge constraints on a Hamiltonian formulation of GR.

[44] This objection is our rendering of a point put to us forcefully and patiently by Carlo Rovelli.

is not witnessed. That is, the dichotomy between Newtonian and tripartite paradigms is not (or need not be) one concerning the fundamental nature of gravity, but one concerned with the best model of a specific situation, and hence where one sets the bar for experimental observation of quantum behavior.

Finally, the proponent of the Newtonian model that we are imagining can articulate clearly the kind of experiment in which they would agree that the quantum nature of gravity was observed, maybe one in which no mere potential could do the job: for instance, the kind of graviton decoherence experiment that we have mentioned, as well as perhaps the retarded GIE protocols discussed by Christodoulou et al. (2022) that we have already referenced, or perhaps some observation of the different proper times read by clocks in different, superposed geometries.[45] Such other experiments might, that is, claim epistemic virtue in having set higher the bar for witnessing the quantum behavior of gravity.

Objection: If gravity is necessarily quantum in a full description, which, as we just emphasized, (nearly) all hands agree, isn't that reason to accept the tripartite model, in order to represent the fact?

Reply: Surely it is a reason, but there are other reasons not to accept the model. First, one could adhere to the maxim that empirically equivalent models which include fewer details are to be preferred: Why retain features that are superfluous to an empirically successful idealization? Such a maxim arguably prefers the Newtonian model precisely because it eliminates relativistic causality.

Second, does adopting the tripartite model in fact succeed in representing gravity as quantum? For the model is in the Newtonian regime, in which the only contribution to gravity is the Gaussian constraint, not any dynamical part of the field. (This point is true even in the specific models given by Bose et al. (2017); Christodoulou and Rovelli (2019); and Chen et al. (2022); if the $|\gamma\rangle$ states were not the specific ones given, then the total state would violate the gauge constraint.[46]) In that case, in the tripartite models, the gravitational state is not describing a dynamical quantum mediator in the way intended in the analysis of Section 5.2.1. But then,

---

[45] The latter suggestion was mooted – as we understand him – by Chaslav Bruckner in discussion.

[46] We thank David Wallace for stressing this point in discussions.

the desire to represent the underlying quantum nature of gravity cannot be a reason to adopt the tripartite model, since that model cannot represent it properly after all! And yet, including a quantum state for gravity, even in such a hobbled way, represents *something* of its dynamical nature....

Stepping back from this to-and-fro, the arguments originally based in the physics have surely left the realm of unequivocal reasons for model choice, and degenerated to value judgments: the question of which model to select has come down to a question of whether some part of the formalism alone well represents a physical property such as the quantum nature of gravity. Our point, then, is that there are compelling physical arguments that accrue to both sides, even granting agreement about LEQG, yet the considerations are not of the kind that can be settled empirically, or by more fundamental principles. They rather "influence decisions without specifying what those decisions must be" (Kuhn [1977b, 362]). Which is why we described the positions as "paradigms."

Objection: The analysis given in Section 5.1 leading to the Newtonian model paradigm depends on a specific choice of gauge. But conclusions about the quantum nature of gravity ought to be gauge-independent; put another way, the conclusion that gravity is classical depends on an arbitrary choice.

Reply: There is nothing per se inappropriate about fixing a gauge to construct a model; one does it all the time. And that is exactly what is done here, *as a means* to take the Newtonian limit, as one must to describe the GIE experiment.[47] (In line with the previous reply, one can of course agree that "really" the field is quantum, yet it is most appropriate to construct a gauge fixed model with gauge-relative observables.)

Taking stock: Having shown the failure of various attempts to find common ground to settle which of the Newtonian model and class of tripartite models, explicated in the first part of the section, is the better description of the GIE experiment, we provisionally conclude that they are paradigms in the sense explained. The merits of this conclusion

---

[47] A completely gauge-independent version of the analysis has been put forward in Anastopoulos et al. (2021, Section 3), dismissing such concerns. But note: Even this more general analysis can only identify the Newtonian potential as the relevant constraint remaining in the nonrelativistic limit upon restricting to the ADM gauge. We do not see this as a bug, but rather as a feature of the analysis.

will be clear in the following section, in which we critically assess the implications of a crucial and much discussed paper in the GIE literature, for interpretations of the experiments. Instead of a practical experiment, Belenchia et al. (2018) propose a thought experiment in which gravity putatively has to be treated as quantized mediator in order to avoid a paradox. With this claim, we essentially agree. However, the introduction to the paper can be read to claim that this conclusion further settles the question of the best account of the GIE experiment (something we have also heard suggested informally):

> …all the previous proposals can be accounted for by just considering the (non-local) gravitational potential in the Schrödinger equation describing the two particles, without any reference to dynamical degrees of freedom of the gravitational field. This has led [Anastopoulos and Hu (2018)] to argue that, even if successful in witnessing entanglement, experiments like [Bose et al. (2017), Marletto and Vedral (2017b)] would say nothing about the quantum nature of the gravitational field. In this work we provide a different conclusion by revisiting a gedankenexperiment …. (p. 1)

If that were correct, then there would be no rational indeterminacy after all; so, we are obligated to consider this objection as well. It will take a dedicated section to do so.

6 Witnessing Gravitational Quanta?

Consider two observers, Alice and Bob, a distance D apart as in Figure 6.1. Assume that Alice has already used a Stern–Gerlach apparatus oriented along the z-axis to prepare a massive particle with positive x-spin in a state

$$\frac{1}{\sqrt{2}}(|L\rangle \otimes |\uparrow\rangle + |R\rangle \otimes |\downarrow\rangle), \tag{6.1}$$

where L, R denote the left and right of Alice, a distance $\delta \ll D$ apart. (Assume further that the state preparation was adiabatic: slow enough that it does not become entangled with relevant fields by emitting radiation.) Now suppose that Bob has placed a massive particle into a trap, such that the particle effectively registers no effect of outside systems.

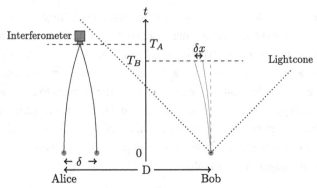

Figure 6.1 Arrangement of the gedankenexperiment of Belenchia et al. (2018)

In these circumstances, were Alice to pass her system (again adiabatically) through a "reversed" Stern–Gerlach apparatus, recombining left and right packets of (6.1) at a single location, then since the particle has remained free of entanglement with other systems, the two spin states would interfere, and Alice would observe a final state of positive x-spin. Let us suppose that she indeed completes this task in a time $T_A > 0$ in the lab frame.

But consider what happens if instead, for a time $T_B > 0$, Bob lets his particle leave the trap and thereby experience the gravitational field of Alice's particle. Because the two packets are at different distances D and $D - \delta$ from Bob's particle, associated with different gravitational forces on Bob's particle, the two particles will entangle: The right (left) packet of Alice's particle becomes correlated with a packet of Bob's slightly more (less) deflected to the left. There then *seems* to arise a clash between "complementarity" and causality, if Alice and Bob are spacelike separated (i.e. $T_A, T_B < D$, with $c = 1$). It seems, that is, that the entanglement of Alice and Bob's particles caused by their interaction entails that the two packets of (6.1) will no longer interfere, and Alice would no longer find a positive x-spin state after passing her system through the "reversed" Stern–Gerlach apparatus.[48] But this would mean that Bob's

---

[48] Note that such an observation will be statistical in nature: that the final state is positive x-spin up requires observing all spin up outcomes in an ensemble of particles prepared by multiple runs of the experiment. Now, even unreliable, probabilistic signaling would violate causality, but this would still require more than a single run of the experiment: a

release of his particle from the trap – which only lasted $T_B < D$ – would have had a faster-than-light effect on Alice's system. Apparently, by deciding whether or not to release his particle, Bob could transmit a bit of data to Alice superluminally, violating causality. If, conversely, Alice's system still did exhibit interference, observed as a final x-spin state, then we could conclude that it remained in a superposition (6.1) of *both* left and right packets during the experiment; yet Bob could measure the deflection of his particle to determine which *one* of the two paths Alice's particle actually took – violating "complementarity"!

Fortunately (for orthodoxy), as Belenchia et al. (2018) show, the apparent clash can be dissolved if the following two more subtle requirements are integrated into the modeling. These we formulate and name as follows:

Minimum length requirement (MLR): Distances in a QG context can only be resolved up to the Planck length L_p. Belenchia et al. argue for this by appealing to the (dominant term of the) vacuum fluctuations of the Riemannian curvature, averaged in a spacetime region of radius R: its correlation function of the (linearized) tensor has a dominant term of form L_P/R^3; integrating the geodesic equation over time R yields an estimated fluctuation in the relative position of two bodies of form $\delta x \sim L_P$. However, note that there is not only one possible argument that Planck length is a minimal distance in QG (as discussed further in subsequent paragraphs), and that the particular value of the Planck length does not do much work in any case.

This requirement can be coupled with the fact that the spatial displacement of Alice's wavepackets leads to an effective (static) quadrupole moment (the dominant gravitational effect since gravity features no dipole radiation). Since the gravitational field difference associated to the effective gravitational quadrupole Q_A is Q_A/D^4, using $m_B \ddot{x} = F_{\text{Newton}}$ it can be estimated that Bob's particle of mass m_B will have been

---

single positive x-spin measurement gives no information about whether Alice's particle is in such a state or in the mixture of $|\uparrow\rangle$ and $|\downarrow\rangle$ expected if it entangles with Bob's particle. To maintain the strictest constraints on causality, Alice and Bob should remain spacelike for the whole series of runs; but if we put to one side the possibility of past runs influencing the outcomes of future runs, we only require that Alice and Bob remain spacelike during each run, as assumed here.

displaced by an amount $\frac{Q_A}{D^4}T_B^2$ after time T_B.[49] But MLR means that this distance has to be greater than L_P for Bob to resolve the path of Alice's particle: that is, it follows that for Bob to obtain "which-path" information about Alice's particle, we must have $\frac{Q_A}{D^4}T_B^2 > L_P$, or $Q_A T_B^2 > D^4$ in $L_P = 1$ units – what we dub the MLR – fulfilled.

Quantized radiation requirement (QRR): When Alice recombines the wavepackets of her particle during time T_A, gravitational radiation is emitted in quanta of frequency $f \sim 1/T_A$. Now, the total energy radiated is given as $E \sim \left(\frac{Q_A}{T_A^3}\right)^2 T_A$.[50] Given the standard de Broglie energy-frequency relation $E = Nf$, it follows that the number of quanta emitted will be $N \sim (\frac{Q_A}{T_A^2})^2$. However, if *any* quanta at all are emitted, then Alice's particle will become entangled with the gravitational field, and coherence will again be lost: So, we must have $N < 1$ if she observes interference. In other words, the requirement $Q_A < T_A^2$ – the QRR – needs to be fulfilled if Alice's particle is to remain in a pure state. (This inequality is so dubbed because of the derivation presented here, but note that later we will ask whether the QRR could be derived without the assumption of quantized gravitational radiation.)[51]

If the localization and radiation requirements are appropriately taken into account, it can easily be shown that there is indeed no paradox, that is, no clash between causality (no superluminal signalling) and complementarity. Consider two cases:

$Q_A < T_A^2$: The QRR is explicitly fulfilled so that Alice can recohere her packets, but the MLR is not fulfilled and Bob cannot measure the path taken by Alice's particle: in order for Alice and Bob's procedures to be spacelike $T_A, T_B < D$, so $T_A^2 T_B^2 < D^4$, and in the current case $Q_A T_B^2 < D^4$, contrary to the MLR. Since it is impossible for both Alice

---

[49] Using the integration estimate $\ddot{x} \sim \delta x T_B^2$.

[50] See Rovenchak and Krynytskyi (2018), Section 3, for a derivation of this quadrupole formula. Note that this radiation is in the form of physical, transverse gravitons. In contrast, the quadrupole field relevant to the MLR is due to the gauge, Gaussian part of the field.

[51] In principle, this radiation could also be measured by Bob at a lightlike separation for its emission, further entangling with him. However, he could not also obtain which-path information, since that would collapse Alice's particle into a single packet, preventing recombination, the source of this radiation.

and Bob to perform their procedures, causality and complementarity will not clash.

$T_A^2 < Q_A$: The QRR is explicitly violated so that Alice's particle is entangled with the gravitational radiation emitted when Alice attempts to recohere her particle, thwarting that attempt. On the other hand, for a suitably large value of Q_A it is possible for the MLR to be satisfied. Again, since we do not have both which-path information and recoherence, there is no clash of causality and complementarity. (In fact, the MLR is that $Q_A T_B^2 / D^4 > 1$; or since $0 < T_B^2 / D^2 < 1$, that $Q_A > D^2 -$ a stricter condition since $D^2 > T_A^2$.)

To sum up, successful completion of the experiment requires that both MLR and QRR are satisfied, but as the analysis of Belenchia et al. (2018) shows, on the assumption of perturbatively quantized linearized gravity – which we will call the "graviton hypothesis" – they cannot be simultaneously satisfied. But if the successful completion of the experiment is thus ruled out, then no clash between causality and complementarity arises. However, the analysis in fact only shows that the graviton hypothesis is sufficient to avoid the clash of causality and complementarity, not that it is necessary (as some of their statements suggest).[52]

But could the case for MLR and QRR really be made without appeal to gravitons? Clearly, this question (in particular with regard to QRR which concerns transverse, i.e. dynamically propagating, gravitons) is important in evaluating the role this thought experiment can play in defending a claim that the experiments with gravcats supply a tabletop witness of the quantum nature of gravity.

On MLR: The derivation of the MLR à la Belenchia et al. operates via a minimal length argument that adheres to vacuum fluctuations. Notably, though, there are a number of arguments for a minimal length

---

[52] Ávila et al. (2022) argue that the assumption of quanta is not even sufficient to evade paradox: Under the assumption that there is not one particle trap on Bob's side but a sufficiently high number N of particles traps, one can obtain decoherence arbitrarily fast by releasing all particles at once – even if the release of a single particle would not lead to decoherence. The more rigorous treatment of the experimental setup by Danielson et al. (2022) – which basically precisifies the "back-on-the-envelope" of Belenchia et al. (2018) presented so far – at the same time would seem to resolve a version of this extended paradox.

(see Hossenfelder [2013]), many of which do *not* make recourse to a quantization of gravity at all. Consider, for instance, the estimate on distance measurement limitations originally going back to Salecker and Wigner (1997) (here presented following Hossenfelder [2013, Section 3.1.3]): If the position of a nonrelativistic clock is known up to Δx, then by the Heisenberg uncertainty relation its velocity can only be known up to $\Delta v = \frac{1}{2M\Delta x}$. Say the clock is a light clock, and time is measured by a photon going back and forth between two mirrors separated by the distance D. Within one clock period $T = 2D$ (c is set to 1 again), the clock thus has an uncertainty in position of $\Delta x + \frac{T}{2M\Delta x}$. The minimal value for Δx (found from a simple variation exercise) is $\Delta x_{min} = \sqrt{T/M} = \sqrt{(2D)/M}$. Requiring a minimal operational length of the clock larger than its Schwarzschild radius (otherwise the clock would be gravitationally unstable), leads to $\Delta x_{min} > \sqrt{4G} \sim l_P$. Again, no quantization assumption about the gravitational field itself has been used to arrive at this estimate! The fact that vacuum fluctuations are not needed to establish MLR shows that there is something misleading about Belenchia et al.'s statement that "both vacuum fluctuations of the electromagnetic field and the quantization of electromagnetic radiation [and, analogously, in the gravitational case] were *essential* for obtaining this consistency [between complementarity and causality]" (p. 5, our stress).

On QRR: As we saw, the condition $Q_A < D^2 < T_A^2$ – and no weaker condition such as $Q_A < T_A^2 + |C|$ – must be satisfied for Alice's particle state to remain coherent. To illustrate this point: One might, for instance, motivate a bound from simply requiring the radiated classical quadrupole energy to be negligible, that is, $E \sim \left(\frac{Q_A}{T_A^3}\right)^2 T_A \ll 1$ in order to ultimately exclude *any* kind of interaction caused by classical radiation linked to Alice's closing of her quadrupole. This would entail a "radiation requirement" of form $Q_A \ll T_A^{5/2}$ (call this CRR) which is a weaker bound than that from QRR for $T_A > 1$ and a stronger bound for $T_A < 1$.[53] The interesting case is that of $T_A < 1$: Then, the CRR can be violated *while* the QRR is fulfilled. So, if the violation of CRR can really be taken to entail that there is relevant interaction caused by

---

[53] Concretely, $T_A^2 < T_A^{5/2}$ for $T_A > 1$, and $T_A^2 > T_A^{5/2}$ for $T_A < 1$.

the closure of the quadrupole and thus a decoherence effect for Alice's spin-state through any mechanism *whatsoever*, the paradox would nevertheless remain: For $T_A < 1$, violation of the CRR would always go along with a violation of the MLR, as the violation of the CRR is accompanied by a fulfillment of the QRR (which in turns implies a violation of the MLR, as shown before).[54] Thus, if the only grounds for the QRR lie in the postulate of radiation quanta, there is quite a straightforward sense in which it requires gravitons, and hence yields a resolution of the paradox that is due to gravitons, as the authors claim (even if the MLR does not require the graviton hypothesis).

However, this claim has been questioned by Anastopoulos et al. (2021, p. 22):

> ...a close reading of the argument [of Belenchia et al.] shows that vacuum fluctuations need not be quantum, and that the restoration of quantum complementarity only requires a decoherence mechanism—spontaneous emission of discrete quanta being only one of possible scenarios. Hence, the arguments of [Belenchia et al.] do not rule out theories in which gravity is treated as a classical stochastic field that causes decoherence to quantum matter, which are properties that any mathematically consistent quantum-to-classical coupling must have, anyway.

First of all, Belenchia et al. (2018, p. 6) agree with the point made earlier that the MLR can be realized without quantisation – and thus they should be happy to grant MLR in any case. So if one buys into the setup of the argument as presented so far, the central issue is this: To what extent does the need for a decoherence mechanism establish a requirement like QRR (or one at least as strong as QRR for $T_A > D$)? Moreover, the account of a purely classical mechanism would have to explain in what sense the naive classical account used for deriving the CRR earlier goes wrong. In any case, we do not know of any such explicit account, and it is difficult to envision how it would go.

The immediate upshot of our critical rehearsal of Belenchia et al. (2018) is as follows then: In the end, in order to resolve the paradox noted

---

[54] More precisely, fulfillment of MLR, violation of CRR and fulfillment of QRR require (i) $Q_A T_B^2 > D^4$, (ii) $Q_A > T_A^{5/2}$ and (iii) $Q_A < T_A^2$, respectively. For $T_A < 1$, (ii) and (iii) combine to $T_A^2 > Q_A > T_A^{5/2}$. But $Q_A < T_A^2$ implies – given that $T_A, T_B < D - Q_A T_B^2 < D^4$, that is, the violation of MLR.

by Belenchia et al. (2018), an explicit viable alternative has to be offered to that of assuming gravitons, that is, to assuming that (i) gravitational radiation comes in (some form of) quanta, and that (ii) this "quantizing" of gravitational radiation is aptly described in the terms provided by perturbatively quantized linearized gravity. In fact, Rydving et al. (2021) have proposed such an alternative by postulating an in principle minimal resolution distance of l_P. Not only is the MLR thus fulfilled, but also a bound sharper than the QRR can be found which together resolve the paradox (we saw that an alternative bound to the QRR has to be at least as strict as the QRR itself); this substitute bound for QRR is obtained by requiring the resolution in Alice's interference measurement to be greater than the Planck length. Whether this proposal is viable then depends on whether a brute postulate of minimal resolution is acceptable (see Großardt (2021) on this question).

But how does the Belenchia et al. (2018) experiment bear on our assessment of the GIE experiment? In particular, having articulated and distinguished between the Newtonian model and tripartite paradigms' analyses of the GIE experiment, we may ask: Does the new thought experiment lead the proponent of the Newtonian model to a new conclusion? Does it help the proponent of the tripartite paradigm?

The Newtonian model proponent is not moved: Notably, the decisive regime for the paradox is a (special) relativistic one, whereas, for the proponent of the Newtonian model, it is a nonrelativistic regime. Recall that the proponent of the Newtonian model asks which gravitational degrees of freedom (in the precise sense discussed earlier) of the (3+1) general relativistic theory still remain relevant in the fully nonrelativistic regime of the GIE experiment, and answers that there are none: The interaction is mediated entirely via the Newtonian potential qua Gauss constraint. There is thus no reason for the proponent of the Newtonian model paradigm to rethink her analysis in the face of a thought experiment unconcerned with the nonrelativistic regime. In short, while Belenchia et al. (2018) show that the tripartite paradigm will resolve the given paradox, this does not imply that it must also be imported to the GIE experiments; our claim that they are "paradigms" stands.[55]

---

[55] We thank Flaminia Giacomini for discussions of the import of the paradox.

If the thought experiment is thus silent regarding the Newtonian model, what are the stakes for the proponent of the tripartite paradigm? First, note that the tripartite paradigm proponent already takes for granted that gravity is quantum in the relativistic regime, having contented herself with the quantum nature of gravity already (by hypothesis) on show in the GIE experiment in the nonrelativistic regime. (Admittedly, the theoretical argument may, however, boost her confidence in that the theoretical picture of gravity should at this scale include gravitons.) On the other hand, the experiment is in a new regime. Let us then, finally, consider how the Belenchia et al. thought experiment could indeed lead to an experiment in the relativistic regime that would provide a tabletop quantum gravity witness of gravitons whatever one's paradigm for GIE. Suppose that the experiment were carried out as described *except* that Alice attempts to recombine her packets quickly enough that $T_A^2 \lesssim Q_A$. The analysis indicates that she will fail because the quadrupole of the packets emits a single graviton, and so ends up in a mixed state. In this modified experiment, then, the absence of interference demonstrates the existence of gravitons. (Conversely, if she recombines them slowly enough that $T_A^2 \gtrsim Q_A$, Bob's failure to determine which path the particle took could be taken as evidence for a minimum length.) Indeed, as we mentioned in footnote 19, it has been proposed that the quantum nature of gravity might be observed in much this way: "gravitationally induced *decoherence*," in contrast with gravitationally induced entanglement.

7 Making Gravity Quantum: Control versus Witness Traditions

Let us take stock of the ground we have covered so far. We presented (Section 4) a new class of proposed experiments relevant to quantum gravity phenomenology – GIE experiments – in which a tabletop pair of "gravcats" promise to probe the quantum nature of the gravitational coupling between them. We were interested, that is, in articulating the payoff of actually performing such experiments within the not-so-distant future. So, What is that payoff?

We considered (Section 4.1) one possible answer, inspired by remarks of many of the actors involved within the quantum gravity phenomenology community: that these experiments would provide "witnesses" of gravity's ultimately quantum nature. We contextualized such a view as being in a *witness tradition* begun by the Page and Geilker (1981) experiment (Section 3.2), understood as attempting to provide evidence of QG. From this perspective, the newly proposed class of experiments constitutes an advance within that tradition, which emphasizes the achievement of ever new kinds of *evidential support* for QG.

We then saw (Section 5), however, that, given our current best physics, the question of whether successful GIE experiments witness the quantum nature is paradigm dependent; our first main philosophical point. In the tripartite paradigm, we saw exactly the sense in which the proposed experiments would provide a novel and subtle kind of witness for gravity's ultimately quantum nature: the payoff apparently identified. Yet, within the Newtonian paradigm, the sense in which the witness provided is one that concerns the quantum nature of *gravity* is muddied at best. So, whether a positive GIE experiment would be a successful advance in the witness tradition is paradigm dependent. To that extent, the value of performing such a difficult experiment might be questioned.

But here's another possible answer to the question of the payoff of the experiment; one which might satisfy regardless of one's paradigm, because it depends only on the broadly shared commitment to LEQG, not on what the GIE experiments may or may not witness according to one's preferred model. Maybe the value of the experiment lies simply in empirically confirming LEQG.

For instance, Wallace (2022) describes empirical applications of the theory to various phenomena: quantum systems in external gravitational fields, self-gravitating systems including stellar evolution, the origin of the CMB, and the Page and Geilker experiment. Since these fall in different physical (sub)regimes, he argues that they amount to successful tests of LEQG in each regime, which collectively support the theory in its full "low-energy" regime (Section 9). He describes these successful applications as theory "confirmation," and indeed they are from a strict hypothetico-deductive point of view (equally, for Popper, failed attempts at refutation).

However, the majority of the applications that Wallace discusses are in regimes in which very little is at stake for LEQG, understood specifically as quantized linear gravity: There is little chance of its refutation, or of a crucial experiment selecting some alternative to it. In particular, almost all of the applications fall in the semiclassical regime, in which the empirical risk to LEQG is small. (Recall from our mentions of Sakharov's induced gravity, that SCG is entailed even by much weaker conditions than quantized linear gravity.) So insofar as LEQG is tested, what is most at risk of refutation is the theory of the material part of the system – a star for instance – rather than the quantum description of the gravitational field itself. But in a more realistic epistemology than hypothetico-deductivism, small risk of refutation means low confirmation, so talk of *confirmation*, while strictly correct, risks overselling the point: Any increase in credence for LEQG is negligible. (To be fair, Wallace's primary point is that LEQG actually makes correct empirical predictions, while the claim about confirmation is secondary.)

A positive outcome in the GIE experiments would (like the CMB data) slightly improve the confirmation of the quantum aspect of gravity, by ruling out SCG as a competitor. However, even in this case, the degree of confirmation is tiny, because SCG is not generally regarded as a serious alternative. So the confirmation story cannot do full justice to any perceived deep significance of the experiments with gravcats. Indeed, this conclusion is foreshadowed by the similar analysis given in Section 3.2. Recall that there, too, it was difficult to articulate the virtues of the experiment in terms of marginal theory confirmation. And we framed their experiment as the first in a witness tradition because the authors themselves chose to emphasize, not confirmation, but empirical access to gravity *being a certain way*: quantum.

Though neither witnessing nor confirmation adequately captures the import of the proposed experiments, we suggest that there is distinct, though compatible, experimental tradition in which to understand their payoff: simply to obtain the *ability* to perform them! That is, to make gravity interestingly quantum in a new way in the laboratory, in the tradition of the COW experiment (Section 3.1). This alternative payoff for the experiments can be accepted within *either* paradigm – the New-tonian as well as the tripartite paradigm – even if they disagree about the interesting way in which gravity is made quantum. For this reason, articulating this second, *control tradition* is essential for our second main philosophical goal: explaining the value of performing the experiments, in a second, complementary, but less paradigm-dependent way.

To begin, consider the immediate upshot of our being able to perform GIE experiments. Even from their naive presentation discussed earlier, we would have controlled the gravitational coupling between two grav-cats in coherent superposition. In the tripartite paradigm, this is (further) to say that we have controlled the quantum state of the gravitational subsystem that mediates entanglement between the two gravcats. On the Newtonian model paradigm, by contrast, we might rather say that we have controlled the constraint part of the gravitational gauge coupling between quantum systems. Regardless of paradigm choice then, we would have achieved "control" in a new regime involving both gravity and quantum mechanics. This situation is comparable with demonstrations that ever bigger macromolecules can remain in superposition for a non-negligible time: Despite some remaining Bohrians, near consensus has it that some form of quantum universalism holds. It's also comparable with increasingly careful Bell-violation experiments. Finally, compare the enthusiasm for GIE experiments with that extended to COW in the history of experimentation within quantum gravity phenomenology: In a nutshell, it is *doing* something not yet done with gravitating quantum material systems, or doing something that until now could not have been done in quite so careful a way with the same. Thus, we can situate the proposed experiments not only as contributing to the witness tradition (according to one paradigm), but, like the earlier COW experiment, also as contributing to the vanguard on an orthogonal tradition of experimentation within quantum gravity phenomenology: a tradition of increasingly sophisticated control over the self-gravitational properties of quantum material systems. In this tradition, the goal is ever to make gravity, so far as concerns quantum material systems in the lab, itself increasingly expansively quantum.

Our point is that recognizing the control aspect of a GIE experiment is crucial to appreciating its significance (moreover, in a relatively paradigm-independent way); of course, within the tripartite paradigm it also has significance as a witness of QG. By further placing it in control and witness *traditions*, we both highlight its commonality with the aims of the earlier COW, and Page and Geilker experiments, respectively;

and stress how it will (at least in the tripartite paradigm) be a significant advance in both regards – though of course leaving much room for further advances!

Our language has been suggestive of Hacking's (1984; 1992) entity realism (and more loosely, Latour [1987]). We haven't fully embraced Hacking's view, since we take the experiments to be potentially evidential for LEQG; while for Hacking performing an experiment only provides grounds for the components of the experimental setup itself, since we have to rely on them to trust that the apparatus functions correctly. In his example, the PEGGY2 experiment does not provide evidence for the weak neutral currents it supposedly detects, but for the electrons that must be reliably "sprayed" for the device to work. But like Hacking, our emphasis is on achievement of know-how, or novel practical means of engaging with fundamental physics as a matter of researchers' practice, and specifically in a laboratory setting through managing to isolate delicate systems from environmental interactions. And indeed, the hypothesized outcomes of the proposed experiments will require extraordinary new success in isolating unusually delicate systems of a kind and in a way that we so far do not know how – something very exciting to anticipate, regardless of one's paradigm.[56]

Aside from the independence of the control tradition analysis from the relevant paradigm choice, we note another advantage: that, in the context of a background theory like LEQG, it makes better sense of the epistemic significance of different "regimes" than we saw the witness tradition do. (Though we emphasize that the idea of control is relatively independent of such a background theory.) With the help of the theory, we can distinguish between different characteristic physical regimes so that the value of performing the GIE experiments can be spelled out in terms of those different regimes.

That is, first: It is compatible with the control tradition that a theorist may help themselves, in their analysis of the experiments, to the interpretive package provided by their favored theories, for example, LEQG. In such terms, the control tradition then asks for control in ever new regimes, but this demand presupposes a theoretical framework in which regimes

---

[56] Hacking likely would not approve of "control": see his (1988).

can be distinguished; it is the very embrace of LEQG that provides the principled means of distinguishing the physical regimes, in each of which we strive to achieve increased control. And so, taxonomies of noteworthy regimes like that provided by Wallace (2022) are exactly what is needed for assessing the extent of our control over the quantum nature of gravity *given LEQG,* and for identifying new regimes in which we should seek control – such as what he calls the "coherent-perturbative nonrelativistic" regime, probed by GIE.

In the witness tradition, it is trickier to see how distinguishing regimes help us to understand the value of different experiments. After all (for a given paradigm), either we have witnessed the quantum nature of gravity in an experiment or we have not. If one succeeded in the first place, why would it be useful to go on and witness QG in another way? Of course, as we have covered, theory confirmation (however marginal) is a reason to perform a witness experiment in another regime. But we see no added epistemic value simply in having witnessed *again* (albeit differently). In other words, we claim that, in the framework set out by Wallace, one can best understand the point of such experiments as GIE from within the control, not witness, tradition.

True enough, the significance of regimes with respect to witnessing is the possibility that different paradigms could set the bar for what counted as observing the quantum nature of gravity in different regimes. For instance, one way to understand the difference between the tripartite and Newtonian paradigms is that the former says that one can witness the quantum nature of gravity in the nonrelativistic regime, while the latter requires the relativistic regime. But note that this is ultimately a point about how to conceptualize the witnessing debate, not a point about the value of achieving a witness by means of some novel experiment.

Returning to the main thread of this subsection, the upshot of our discussion is that there is a second payoff of the new experiments, if successful: They would affirm that we are able to *make gravity interestingly quantum on the tabletop* in a manner that exceeds our capabilities to do so thus far. And this is something that one can accept regardless of one's paradigm, so regardless of one's view on witnessing. We thus see an unequivocal sense in which the experiment stands to teach us something important and new.

But why, then, is this not the standard line? Why, that is, do we see the primary actors involved as participating rather in the witness tradition, and where, their having embraced the tripartite paradigm, the testing-in-different-regimes benefit of the experiments is rendered merely a satisfying afterthought? We submit the following explanation. Recall the three views of SCG. It strikes us that the witness tradition is very natural (though, we stress, not logically required) on view 1, contrary to view 2: There is an explicit candidate theory standing against taking SCG as fundamental. Refuting view 2, with view 1 the alternative, would seem just to amount to a witness of gravity's being quantum, according to the latter. Meanwhile, the control tradition is very natural on view 3, contrary to view 2: where, to the extent that there is some such micro-physics giving rise to SCG, we should like to learn about it by whatever means avail themselves. But view 3 is just not that popular compared to view 1 – absent strong explicit rivals to LEQG gathered under the same net, view 3 seems more like a pedantic reminder of the vastness of logical possibility space (so far as concerns means of saving the phenomena). Hence, we offer a genealogical explanation of the present circumstance: In contemplating bringing together quantum theory and gravity, SCG was offered. But it was appealing to understand SCG itself, if/since it could not be fundamental (view 2), specifically in terms of LEQG (view 1). So, experiments pursued were pursued in an effort to put forth increasingly sophisticated witnesses of the quantum nature of that specific underlying theory.

That is, once again, recognition of the control tradition is necessary to fully articulate the significance of tabletop quantum gravity experiments.

Let us use a very recent proposal by Howl et al. (2021) to demonstrate further the importance of recognizing the control tradition: Arguments focused too narrowly on a question of tabletop quantum gravity witness will miss otherwise obvious virtues of novel experimental protocols. Howl et al. present their tabletop quantum gravity experiment in terms of a witness payoff akin to that for the GIE experiments, but without entanglement as its central mechanism. They presuppose an analogue to the tripartite paradigm for the GIE experiments, now taking a Bose–Einstein condensate's (BEC) transition from a Gaussian to a non-Gaussian state in the presence of a gravitational interaction as indicating that the gravitational mediator is quantum in nature.[57] At its heart, the proposal is based on the insights that (i) non-Gaussianity is measurable in a BEC, that (ii) a quantum interaction between the BEC atoms is needed to bring about an overall non-Gaussian state, and that (iii) a BEC can be controlled in such a way that no other interaction but the gravitational one effectively occurs between the BEC atoms. The upshot is a Gravitationally Induced Non-Gausianity (one might say "GING") experiment, in parallel with the program of GIE experiments.

Notably, the experiment yet again cannot make a difference in the in-principle quarrel between the Newtonian model paradigm and the tripartite models paradigm: As in the case of our analysis of the GIE experiments, whether the GING experiment allows for actually testing the quantum nature of gravity is contingent on choices on which reasonable people may disagree. In fact, the analysis of gravity in the GING experiment is very similar to that in the GIE experiments; so it might seem, from that point of view, to offer little advance over the latter. Why not concentrate our efforts, then, on just one? One immediate response comes from a practical perspective: The more cutting-edge scenarios that are available in which we are close to witnessing QG (in the tripartite paradigm), the more likely it is that there is a technical break-through that would make possible our witnessing QG — and, indeed, a GING experiment may well be much easier to successfully execute. Another possible response is more subtle: Perhaps one is interested in regime-specific witnesses of QG, in witnessing the quantum nature of gravity *within each notable regime*. However, such a response depends delicately on the way we define regimes: For instance, although the use of BECs arguably implies a different regime, both GIE and GING experiments fall in the same perturbative-coherent-nonrelativistic regime according to Wallace's (2022, Table 2) taxonomy.

Particularly because of this ambiguity, the proposal of seeking *witnesses* across regimes strikes us as weakly motivated: Without careful

---

[57] In what seems to us an instance of multiple invention, Haine (2021) proposed a distinct BEC experiment intended to detect quantum effects of the gravitational interaction. Notably, unlike Howl et al. (2021), Haine (2021, Section 5) explicitly contrasts the merits of their proposal with the (in their view) limited focus of analyses in the GIE experiments only on achieving an entanglement witness.

inspection of one's taxonomy of regimes, it is unclear what virtues there are to achieving a witness of the underlying quantum nature of gravity within each regime, so individuated. One either has witnessed QG or has not (relative to a choice of paradigm), no matter the regime in which one has found it. In contrast, in the control tradition, every further experimental intervention in every further regime is an advance. In other words, the hopes of the experiments in the witness tradition would seem to rise and fall together; the virtues of diversity of experiment are lost. By contrast, in the control tradition the mixed bag of "advantages" and "disadvantages," claimed or otherwise (cf. Howl et al. [2021]), of the GING experiment over the GIE experiment is reinterpreted as an explicit discussion of the differences in the kind of access to or control of the gravitational interaction in various quantum experiments. Indeed, differences in ways of controlling make the different experiments well worth performing!

8 Concluding Remarks: Taking Stock of Quantum Gravity Phenomenology

According to lore in the philosophy of QG, the problem of QG is (very nearly) purely one for the theoreticians. It is just too difficult to hope for discriminating signatures of QG in data, because the relevant empirical regimes far exceed our capacities for experimentation (in high-energy physics) or direct detection (in astrophysics and cosmology). But this lore is misleading of fundamental physics practice today. In recent decades, and to wide acclaim in the surrounding discipline, a range of empirical testing strategies have been pursued within the arena of quantum gravity phenomenology, as a proposed means of gaining significant, increased empirical traction on the problem of QG.

Our focus throughout this Element has been on disentangling threads in the interpretation of experiments within one such emerging empirical testing strategy: tabletop quantum gravity. Our impetus was evident disagreement among physicists involved in tabletop quantum gravity concerning a new class of proposed experiments that are otherwise by and large constitutive of the new subject.[58] Hopefully, we have clarified

---

[58] In fact, there is, arguably, another class of experiments in tabletop quantum gravity that we ignored: experiments attempting to demonstrate indefinite causal order on the

that such disagreement reflects something more subtle and foundation-ally noteworthy than a simple confusion by some parties to the dispute. On one hand, reflecting on the physical legitimacy of the Newtonian model counsels against any claim that the GIE experiments would, by hypothesis, provide a witness of the quantum nature of gravity, given the interpretational framework provided by the quantization of gauge theories. On the other hand, reflecting on the tripartite models counsels trust in the very same claims, given the treatment of gravity as a mediating subsystem. Thus, one has the following moral: How we assess the empirical scope of tabletop quantum gravity experiments is a much more intricate affair than may be at first thought, and crucially depends on matters of physical interpretation that are not settled by rational argument.

Such intricacy is perhaps not surprising in the philosophy of experimentation. Disagreements over the stakes of experiments or observations are almost certainly to be found in disagreements over interpretations of the naive descriptions assigned to the relevant experimental setups, given the sum total of the lessons provided about the world by our current best physics. And just so: Our explanation of the disagreement over the claim of quantum gravity witness in the new GIE experiments in table-top quantum gravity was sensitive to paradigmatic disagreement about the experiments, understood by current lights. Still, it is interesting to see how the general thesis shows up, in particular, where: On the one hand, most theorists involved already endorse LEQG en route to developing a future high-energy theory of QG; while on the other, there are different modeling strategies compatible with that endorsement (not to mention more agnostic approaches available to merging together our current corpus of physics than an embrace of LEQG). And so, there are diverging identifications within quantum gravity phenomenology as to what properly amounts to our bringing to bear "our current best physics" on the

---

tabletop, for example, (Rubino et al. 2017). Now, there is a sense in which conformal structure is a more profound component of the spacetime metric in GR, and so perhaps experiments focused on producing superposition states of conformal structure ought to be distinguished from the rest. It seems dubious to us that there are, indeed, ultimately satisfying reasons to treat conformal structure in this privileged way, however. But even so, we hope that the careful attention we have given to the question of the relationship between QG witness and metric superposition states by means of GIE experiments is informative in thinking through possible upshots of these other proposed experiments.

analysis of any laboratory gravitational experiment with quantum matter probes. From this perspective, the new class of experiments is arguably a first arena where ambiguities at the community level over what to identify as our current best physics spell fierce disagreement among those otherwise allied in an empirical-first pursuit of a high-energy theory of QG.

Nevertheless, as we saw in the previous section, there are means of packaging the empirical stakes of such tabletop quantum gravity experiments that are largely immune to possible foci of disagreement over the matters of interpretation. Great care must therefore be given to whether bottom-line endorsements of new experiments on the horizon are defended on argumentative grounds sensitive to interpretation (in the sense intended here), or on argumentative grounds that are separable from such interpretive commitments. In particular, the tradition of discussing experiments in tabletop quantum gravity in terms of witnesses of the quantum nature of gravity is rife with troubles born by such sensitivity, while the tradition that emphasizes control over the nature of the (typically presumed quantum) gravitational coupling between quantum matter probes seems more promising for cashing out the stakes of such experiments. In certain contexts, this tradition would seem to reduce to talk about grappling with the world across various distinct physical regimes, but we have argued that such a position is derivative of further (common) assumptions about the relevance of LEQG, which need not be embraced to appreciate the same experiments.

From all of this, two meta-philosophical lessons on future work immediately follow: First, philosophy of *frontier fundamental physics* (especially philosophy of QG) usually focuses on matters of theory and is little interested in experiments.[59] We take our work to show that in

---

[59] Welcome recent exceptions are, for instance, works on the status of experiments and simulations in the context of cutting-edge astrophysics relating to outstanding questions in fundamental physics (see, for instance, Anderl [2018]; Gueguen [2020]) or regarding gravitational Hawking radiation (Dardashti et al. [2017]; Crowther et al. [2021]), as well as work on exploratory experimentation in high-energy/beyond-the-standard-model particle physics (Karaca [2017]; Beauchemin [2020]). But we stress that these exceptions, including now our own, do not amount to a victory: There is simply an enormous (and growing) body of work connecting frontier moves in theoretical fundamental physics to the empirical world, which has barely been grazed by philosophers.

times of crisis, clarifications of what we mean and intend with certain experiments can be just as philosophically exciting (and important) as what we normally discuss with relation to pure theory. Second, with our Element, we hope to have provided a case for a philosophy *in* physics (in the style of what Pradeu et al. [2021] call "philosophy in science") that engages with an ongoing debate in physics in real time and not from hindsight, and which operates (at least) with the (side) goal of being of actual use to the practitioner. In fact, what one could witness, or so we hope, is a prime example of how an outstanding controversy between physicists can at times *only* be settled by making recourse to the philosophers' toolbox. It is in these two respects that we take it that philosophers of physics can contribute more to the search for a future theory of quantum gravity!

Appendix
A Newton–Cartan Analysis of Gravcats

Here we derive a general expression for differential contributions to the phase factor for a quantum test particle of mass m subject to Newtonian gravity, due locally to the Newton–Cartan spacetime geometry in the neighborhood of the particle. ("Differential" in the sense that the phase is defined relative to a choice of zero-point: the spacetime geometry for which there is understood to be no such additional contribution to the phase.) The following is intended to be analogous to the analysis provided in the (linear, post-Newtonian regime) general relativistic case by Christodoulou and Rovelli (2019), discussed in §5.2.2.

As in their analysis, the basis states of the gravitational field in a gravcat QG witness experiment are assumed to correspond, in the neighborhood of either one of the gravcats, to what may be described classically as the local gravitational field. This field is associated with a self-gravitation effect, as well as an effect due to a source mass (the other gravcat) at some spatial distance that remains approximately fixed (for the short lifetime of the experiment). Following Christodoulou and Rovelli (2019), we assume that in all but one classical configuration of the two gravcats – the one in which they are closest together – the gravitational self-contribution of any one particle to its phase factor is so great as to dominate the effect of the source particle, at least over the lifetime of the experiment. This self-contribution thus defines the zero-point phase contribution, and we only need determine how the state in which the gravcats are closest modifies it to find the differential contribution.[1]

Formally, let $(M, t_a, h^{ab}, \overset{\circ}{\nabla})$ be a classical spacetime: $t_a h^{ab} = 0$ for all $p \in M$ and $\overset{\circ}{\nabla}$ is a flat derivative operator on M compatible with t_a and h^{ab}. For some curve (with endpoints) $\gamma : [s_1, s_2] \rightarrow M$ modeling the worldline of one of the two gravcats – idealized as a test particle of mass m located at its own center of mass – through the duration of

---

[1] That is, the phase contribution is effectively given (up to a multiple) by the total energy *over time* associated to the two gravcats closest together. Cf. Christodoulou and Rovelli (2019).

the experiment, we will assume that $(M, t_a, h^{ab}, \overset{\circ}{\nabla})$ is apt as a model of the geometric features of the experimental setup, at least in a sufficiently small neighborhood of γ (note that which of the two gravcats we pick for this purpose does not matter, due to the symmetries of the experimental setup). Consider two Newton–Cartan models of Newtonian gravity in the neighborhood of γ, defined respectively as (M, t_a, h^{ab}, ∇) and $(M, t_a, h^{ab}, \nabla')$: that is, the models differ at most in their curved derivative operators ∇ and ∇'. Without loss of generality, we may regard these two models as Trautman geometrizations of two models of standard Newtonian gravitation on $(M, t_a, h^{ab}, \overset{\circ}{\nabla})$, corresponding to gravitational potentials ϕ and ϕ' (satisfying Poisson's equation on $(M, t_a, h^{ab}, \overset{\circ}{\nabla})$). In the neighborhood of γ, ϕ and ϕ' may, as noted, be decomposed into a self-gravity term (identical in each case) and a term due to the gravcat's interaction with the other gravcat situated at a spatial distance, fixed for the duration of the experiment. In what follows, we will take $\phi' > \phi$ so that $(M, t_a, h^{ab}, \nabla')$ is the geometrized model of the gravitational field when the other gravcat is nearest. For convenience then, we will speak of ∇' as the "errant" geometry, relative to the standard set by ∇, that is, the (geometrized) gravitational model (M, t_a, h^{ab}, ∇) that constitutes the "lab-average". Assume from hereon that γ is geodesic with respect to ∇, but whether the gravcat traversing γ is understood to survey ∇ or the errant geometry ∇' depends on the configuration of the two gravcats. Since their separation will differ across branches of the joint wavefunction (4.5), one immediately sees that there will be branch-relative differences in spacetime geometry, reflecting branch-relative differences in the local gravitational field about γ.

From the Trautman geometrization lemma, we know that $\nabla = (\overset{\circ}{\nabla}, -t_a t_b h^{cd} \nabla_d \phi)$ and $\nabla' = (\overset{\circ}{\nabla}, -t_a t_b h^{cd} \nabla_d \phi')$,[2] where in each case we exploit the fact that $\overset{\circ}{\nabla}_a \psi = \nabla_a \psi = \nabla'_a \psi$ for any scalar field ψ. From this, one can easily show that, since $\xi^m \nabla_m \xi^a = \mathbf{0}$ for the four-velocity ξ^a of the gravcat (i.e., whose integral curve is γ), then $\xi^m \nabla'_m \xi^a = \xi^m \nabla'_m \xi^a - \xi^m \nabla_m \xi^a = \xi^m \xi^n t_m t_n h^{ab} \nabla_b (\phi' - \phi)$. In other words, the acceleration of ξ^a with respect to ∇' relevant in the one branch of the wavefunction, given that the integral curve γ is geodesic with respect to ∇, is fully

---

[2] This notation is explained in Malament (2012).

determined by the difference in potentials between the one branch and the rest. (Note that this expression is fully general: if $\phi = \phi'$ up to the addition of a constant, then $\nabla = \nabla'$ up to a constant multiple, and so the acceleration of the gravcat that surveys the errant geometry can be seen to vanish necessarily with respect to ∇', given that the integral curve is geodesic with respect to ∇.)

We may now consider the quantity of power that must be input into the gravcat at any point in its trajectory – in the branch in which it surveys the errant geometry ∇' – relative to the zero power that is needed for the same trajectory in the branches whose gravitational field approximates the lab average spacetime geometry (M, t_a, h^{ab}, ∇). Since acceleration is spacelike, we know that there exists[3] a covector u_b such that $h^{ab} u_b = \xi^m \xi^n t_m t_n h^{ab} \nabla_b (\phi' - \phi)$. So, for $Power = (m \cdot u_b) \xi^b = m \cdot (u_b \xi^b)$ ("power = force × velocity"), it follows that

$$Power = m \xi^b (\xi^m \xi^n t_m t_n) \nabla_b (\phi' - \phi). \tag{A.1}$$

It is important to stress that this quantity of power is defined pointwise along γ. What this means is that we may consider the integral of this expression over the curve γ to compute the total power associated with the gravcat's surveying ∇', through the course of its geodesic path in the lab-average spacetime.

Finally, may consider the amount of (virtual) work done on the gravcat surveying ∇' at each instant along γ: the quantity of power multiplied by the "proper"[4] time $t_c \xi^c$ experienced by the particle at that instant.[5] Integrated over the curve γ, this scalar product of power and proper time yields an expression for the total energy input to the gravcat

---

[3] See proposition 4.1.1 and the discussion immediately preceding it in Malament (2012).

[4] Here, "proper" just notes that the scalar quantity is again defined pointwise, that is, as a degenerate tensor; it is a further fact about classical spacetimes that, integrated over a curve between endpoints, proper time elapsed along the curve necessarily agrees with the global time elapsed between the two endpoints.

[5] As observed by Weatherall and Manchak (2014), we may, in effect, regard this quantity of work as the result of an ordinary force field on (M, t_a, h^{ab}, ∇), which acts only on particles there that would survey the errant ∇'. Meanwhile, the force field that acts there on particles that survey ∇ is the 0 tensor, and so if $\nabla' = \nabla$ up to a constant multiple (corresponding to $\phi' = \phi$ up to an additive constant) zero work is put into the particle. Note that, as an upshot, the computed quantity of energy here is no more or less mysterious than that which is put into any charged particle in the presence of a fixed, ambient field, which deflects that charged particle off of its geodesic path in accordance

over the lifetime of the experiment, with respect to (M, t_a, h^{ab}, ∇), for its surveying the errant ∇':

$$Total\ Energy = m \int_\gamma \xi^b (\xi^m \xi^n t_m t_n) \nabla_b (\phi' - \phi)(t_c \xi^c) dS. \qquad (A.2)$$

In the specific case of the experiment, for either choice of gravcat, we may note that – at least to good approximation over the short lifetime of the experiment – the difference $\phi' - \phi$ is symmetric under time translations for the duration. In other words, adopting coordinates appropriate for the experimental setup (for definiteness, considered in the lab-average case), this entire expression evaluates to $m(\phi' - \phi) \cdot T$ where T is the total time elapsed and $(\phi' - \phi)$ is understood as a function solely of spatial coordinates at a point. (This follows from separating out the expression for the total power from the expression for the total proper time elapsed. That the second yields T is trivial; that the first yields $m(\phi' - \phi)$ follows from the time translation symmetry of the difference $\phi' - \phi$.) Moreover, noting that the self-gravity contributions to the potential about γ cancel, it is easy to see that the differential contribution to the phase factor provided by the errant geometry relevant in the one branch relative to the rest, which is there due to the nearer – hence, stronger – mutual gravitational attraction of the two gravcats so geometrized, is identical to that which is calculated in the main text, in what was there dubbed the naive account (see eq. 4.3).

To see that the expression calculated is the correct quantity, consider the following prescription for quantization, based on the work of Bohm et al. (1987) (though we will not be taking their "ontological" interpretation here). They note that if one writes the wavefunction as $\psi = R e^{iS/\hbar}$, with $R = |\psi|$ and S the phase, then the time-dependent Schrödinger equation

$$i\hbar \partial_t \psi = -\frac{\hbar^2}{2m} \vec{\nabla}^2 \psi + V\psi \qquad (A.3)$$

yields for the real and imaginary parts (respectively):

$$\partial_t S + (\vec{\nabla} S)^2 / 2m + V - \hbar^2 \vec{\nabla}^2 R / 2mR = 0, \qquad (A.4)$$

---

with a force law associated with the field-charge pair. This will be important shortly, in promoting this quantity of total energy to a quantum phase factor.

and

$$\partial_t R^2 + \vec{\nabla}(R\vec{\nabla}S)/m = 0. \tag{A.5}$$

Or, in the case that R is a constant with respect to space and time, as for instance in the plane wave that describes a gravcat state, the first reduces to

$$\partial_t S + (\vec{\nabla}S)^2/2m + V = 0, \tag{A.6}$$

which is the Hamilton–Jacobi equation for the corresponding classical particle (recall that $\vec{p} = \vec{\nabla}S$), with S now interpreted as the action. Or, reversing our reasoning to this point, we can take as an equivalent quantization procedure for such systems that one takes the classical action and promotes it ($\times\, i/\hbar$) to the phase of a wavefunction.[6] But the quantity that we just computed for the Newton–Cartan gravcats, namely total energy over time, effectively is the classical action (there is no potential term so that the time integral over the Hamiltonian (total energy) and over the Lagrangian (classical action) are effectively the same), and thus precisely is the phase that we seek.

---

[6] Now, because we also have (A.5), we do not have classical particle mechanics: As is well-known, in Bohmian mechanics the wavefunction determines the velocity of a particle at any point, not (only) its acceleration. Nevertheless, Bohm and Hiley propose understanding the last term in (A.4) as a "quantum potential." All this is of course besides the point in the gravcat system in which R is constant and the term vanishes.

References

Adlam, E. (2022), "Tabletop experiments for quantum gravity are also tests of the interpretation of quantum mechanics," *Foundations of Physics* **52**(5), 115.

Altamirano, N., Corona-Ugalde, P., Mann, R. B. and Zych, M. (2018), "Gravity is not a pairwise local classical channel," *Classical and Quantum Gravity* **35**(14), 145005.

Amelino-Camelia, G. (2013), "Quantum-spacetime phenomenology," *Living Reviews in Relativity* **16**(1), 1–137.

Anastopoulos, C. and Hu, B. L. (2014), "Problems with the Newton–Schrödinger equations," *New Journal of Physics* **16**(8), 085007.

Anastopoulos, C. and Hu, B. L. (2015), "Probing a gravitational cat state," *Classical and Quantum Gravity* **32**(16), 165022.

Anastopoulos, C. and Hu, B. L. (2018), "Comment on 'a spin entanglement witness for quantum gravity' and on 'gravitationally induced entanglement between two massive particles is sufficient evidence of quantum effects in gravity,'" *arXiv preprint arXiv:1804.11315*.

Anastopoulos, C. and Hu, B. L. (2022), "Gravity, quantum fields and quantum information: Problems with classical channel and stochastic theories," *Entropy* **24**(4), 490.

Anastopoulos, C., Lagouvardos, M. and Savvidou, N. (2021), "Gravitational effects in macroscopic quantum systems: A first-principles analysis," *Classical and Quantum Gravity* **38**(15), 28.

Anderl, S. (2018), "Simplicity and simplification in astrophysical modeling," *Philosophy of Science* **85**(5), 819–831.

Ávila, P., Okon, E., Sudarsky, D. and Wiedemann, M. (2022), "Quantum spatial superpositions and the possibility of superluminal signaling," *arXiv preprint arXiv:2204.01190*.

Ballentine, L. E. (1982), "Comment on 'indirect evidence for quantum gravity,' " *Physical Review Letters* **48**(7), 522.

Beauchemin, P.-H. (2020), "Signature-based model-independent searches at the large hadron collider: An experimental strategy aiming at safeness in a theory-dependent way," *Philosophy of Science* **87**(5), 1234–1245.

Belenchia, A., Wald, R. M., Giacomini, F. et al. (2018), "Quantum superposition of massive objects and the quantization of gravity," *Physical Review D* **98**(12), 126009.

Berry, M. (1982), "Wavelength-independent fringe spacing in rainbows from falling neutrons," *Journal of Physics A: Mathematical and General* **15**(8), L385.

Bohm, D., Hiley, B. J. and Kaloyerou, P. N. (1987), "An ontological basis for the quantum theory," *Physics Reports* **144**(6), 321–375.

Bose, S., Mazumdar, A., Morley, G. W. et al. (2017), "Spin entanglement witness for quantum gravity," *Physical Review Letters* **119**(24), 240401.

Bose, S., Mazumdar, A., Schut, M. and Toroš, M. (2022), "Two mechanisms for quantum natured gravitons to entangle masses," *arXiv preprint arXiv:2201.03583*.

Brown, H. R. (1996), 'Bovine metaphysics: Remarks on the significance of the gravitational phase effect in quantum mechanics' R. Clifton, ed., in *Perspectives on quantum reality*, Springer, pp. 183–193.

Burgess, C. P. (2004), "Quantum gravity in everyday life: General relativity as an effective field theory," *Living Reviews in Relativity* **7**(1), 1–56.

Carney, D., Stamp, P. C. and Taylor, J. M. (2019), "Tabletop experiments for quantum gravity: A user's manual," *Classical and Quantum Gravity* **36**(3), 034001.

Chen, L.-Q., Giacomini, F. and Rovelli, C. (2022), "Quantum states of fields for quantum split sources," *arXiv preprint arXiv:2207.10592*.

Christian, J. (1997), "Exactly soluble sector of quantum gravity," *Physical Review D* **56**(8), 4844.

Christodoulou, M., Di Biagio, A., Aspelmeyer, M. et al. (2022), "Locally mediated entanglement through gravity from first principles," *arXiv preprint arXiv:2202.03368*.

Christodoulou, M., Di Biagio, A., Howl, R. and Rovelli, C. (2022), "Gravity entanglement, quantum reference systems, degrees of freedom," *Classical and Quantum Gravity*.

Christodoulou, M. and Rovelli, C. (2019), "On the possibility of laboratory evidence for quantum superposition of geometries," *Physics Letters B* **792**, 64–68.

Colella, R., Overhauser, A. W. and Werner, S. A. (1975), "Observation of gravitationally induced quantum interference," *Physical Review Letters* **34**(23), 1472.

Crowther, K. and Linnemann, N. (2019), "Renormalizability, fundamentality, and a final theory: The role of UV-completion in the search for quantum gravity," *The British Journal for the Philosophy of Science* **70**(2), 377–406.

Crowther, K., Linnemann, N. and Wüthrich, C. (2021), "What we cannot learn from analogue experiments," *Synthese* **198**(16), 3701–3726.

Danielson, D. L., Satishchandran, G. and Wald, R. M. (2022), "Gravitationally mediated entanglement: Newtonian field versus gravitons," *Physical Review D* **105**(8), 086001.

Dardashti, R., Thébault, K. P. Y. and Winsberg, E. (2017), "Confirmation via analogue simulation: What dumb holes could tell us about gravity," *The British Journal for the Philosophy of Science* **68**(1), 55–89.

Dawid, R. (2013), *String theory and the scientific method*, Cambridge University Press.

Deutsch, D. and Marletto, C. (2015), "Constructor theory of information," *Proceedings of the Royal Society A: Mathematical, Physical and Engineering Sciences* **471**(2174), 20140540.

Diósi, L. (1989), "Models for universal reduction of macroscopic quantum fluctuations," *Physical Review A* **40**(3), 1165.

Dürr, D., Goldstein, S. and Zanghi, N. (1992), "Quantum mechanics, randomness, and deterministic reality," *Physics Letters A* **172**(1–2), 6–12.

Ehlers, J. (2019), "Republication of: On the Newtonian limit of Einstein's theory of gravitation," *General Relativity and Gravitation* **51**(12), 1–20.

Elder, J. (2022), "On the 'direct detection' of gravitational waves," *PhilSci-Archive preprint:* philsci-archive.pitt.edu/21944/.

Fragkos, V., Kopp, M. and Pikovski, I. (2022), "On inference of quantization from gravitationally induced entanglement," *arXiv preprint arXiv:2206.00558.*

Franklin, A. D. (2017), "Is seeing believing?: Observation in physics," *Physics in Perspective* **19**(4), 321–423.

Galley, T. D., Giacomini, F. and Selby, J. H. (2022), "A no-go theorem on the nature of the gravitational field beyond quantum theory," *Quantum* **6**, 779.

Ghirardi, G. C., Rimini, A. and Weber, T. (1986), "Unified dynamics for microscopic and macroscopic systems," *Physical review D* **34**(2), 470.

Greenberger, D. M. and Overhauser, A. (1979), "Coherence effects in neutron diffraction and gravity experiments," *Reviews of Modern Physics* **51**(1), 43.

Greenberger, D. M. and Overhauser, A. W. (1980), "The role of gravity in quantum theory," *Scientific American* **242**(5), 66–77.

Großardt, A. (2021), "Comment on 'do Gedankenexperiments compel quantization of gravity,'" *arXiv preprint arXiv:2107.14666.*

Gueguen, M. (2020), "On robustness in cosmological simulations," *Philosophy of Science* **87**(5), 1197–1208.

Hacking, I. (1984), "Experimentation and scientific realism." A. I. Tauber, ed., in *Science and the Quest for Reality*, Springer, pp. 162–181.

Hacking, I. (1988), "Philosophers of experiment," in *PSA: Proceedings of the Biennial Meeting of the Philosophy of Science Association*, Vol. 1988, Cambridge University Press, pp. 147–156.

Hacking, I. (1992), "The self-vindication of the laboratory sciences," in A. Pickering, ed., *Science as Practice and Culture*, Chicago University Press, pp. 29–64.

Haine, S. A. (2021), "Searching for signatures of quantum gravity in quantum gases," *New Journal of Physics* **23**(3), 033020.

Hall, M. J. and Reginatto, M. (2018), "On two recent proposals for witnessing nonclassical gravity," *Journal of Physics A: Mathematical and Theoretical* **51**(8), 085303.

Hartle, J. B. and Horowitz, G. T. (1981), "Ground-state expectation value of the metric in the 1/N or semiclassical approximation to quantum gravity," *Physical Review D* **24**(2), 257.

Hesse, M. B. (2005), *Forces and fields: The concept of action at a distance in the history of physics*, Courier Corporation.

Hossenfelder, S. (2013), "Minimal length scale scenarios for quantum gravity," *Living Reviews in Relativity* **16**(1), 1–90.

Howl, R., Vedral, V., Naik, D. et al. (2021), "Non-gaussianity as a signature of a quantum theory of gravity," *PRX Quantum* **2**(1), 010325.

Hu, B. L. and Verdaguer, E. (2020), *Semiclassical and stochastic gravity: Quantum field effects on curved spacetime*, Cambridge University Press.

Hu, B. L. and Verdaguer, E. (2008), "Stochastic gravity: Theory and applications," *Living Reviews in Relativity* **11**(1), 1–112.

Huggett, N. and Callender, C. (2001), "Why quantize gravity (or any other field for that matter)?," *Philosophy of Science* **68**(S3), S382–S394.

Huggett, N. and Wüthrich, C. (2020), "Out of nowhere: The 'emergence' of spacetime in string theory," *arXiv preprint arXiv:2005.10943*.

Jacobson, T. (1995), "Thermodynamics of spacetime: The Einstein equation of state," *Physical Review Letters* **75**(7), 1260.

Kafri, D., Milburn, G. and Taylor, J. (2015), "Bounds on quantum communication via Newtonian gravity," *New Journal of Physics* **17**(1), 015006.

Kafri, D., Taylor, J. and Milburn, G. (2014), "A classical channel model for gravitational decoherence," *New Journal of Physics* **16**(6), 065020.

Karaca, K. (2017), "A case study in experimental exploration: Exploratory data selection at the large hadron collider," *Synthese* **194**(2), 333–354.

Kleinert, H. (2016), *Particles and quantum fields*, World Scientific.

Kuhn, T. S. (1962), "The structure of scientific revolutions," *International Encyclopedia of Unified Science* **2**(2), XV–172.

Kuhn, T. S. (1977a), *The essential tension*, University of Chicago Press.

Kuhn, T. S. (1977b), "Objectivity, value judgment, and theory choice," in *Arguing about science*, pp. 74–86.

Latour, B. (1987), *Science in action: How to follow scientists and engineers through society*, Harvard University Press.

Lin, H. (2022), "Bayesian epistemology," in E. N. Zalta and U. Nodelman, eds., *The Stanford Encyclopedia of Philosophy*, Metaphysics Research Lab, Stanford University.

Malament, D. B. (1995), "Is Newtonian cosmology really inconsistent?," *Philosophy of Science* **62**(4), 489–510.

Malament, D. B. (2012), *Topics in the foundations of general relativity and Newtonian gravitation theory*, University of Chicago Press.

Mannheim, P. D. (1998), "Classical underpinnings of gravitationally induced quantum interference," *Physical Review A* **57**(2), 1260.

Marletto, C. and Vedral, V. (2017*a*), "Gravitationally induced entanglement between two massive particles is sufficient evidence of quantum effects in gravity," *Physical Review Letters* **119**(24), 240402.

Marletto, C. and Vedral, V. (2017*b*), "Witnessing the quantumness of a system by observing only its classical features," *npj Quantum Information* **3**(1), 1–4.

Marletto, C. and Vedral, V. (2019), "Answers to a few questions regarding the bmv experiment," *arXiv preprint arXiv:1907.08994*.

Marletto, C. and Vedral, V. (2020), "Witnessing nonclassicality beyond quantum theory," *Physical Review D* **102**(8), 086012.

Marshall, W., Simon, C., Penrose, R. and Bouwmeester, D. (2003), "Towards quantum superpositions of a mirror," *Physical Review Letters* **91**(13), 130401.

Marshman, R. J., Mazumdar, A. and Bose, S. (2020), "Locality and entanglement in table-top testing of the quantum nature of linearized gravity," *Physical Review A* **101**(5), 052110.

Misner, C. W., Thorne, K. S. and Wheeler, J. A. (1973), *Gravitation*, Macmillan.

Møller, C. (1962), "Les théories relativistes de la gravitation," *Colloques Internationaux CNRS* **91**(1), 15–29.

Okon, E. and Callender, C. (2011), "Does quantum mechanics clash with the equivalence principle—and does it matter?," *European Journal for Philosophy of Science* **1**(1), 133–145.

Overhauser, A. and Colella, R. (1974), "Experimental test of gravitationally induced quantum interference," *Physical Review Letters* **33**(20), 1237.

Padmanabhan, T. (2014), "General relativity from a thermodynamic perspective," *General Relativity and Gravitation* **46**(3), 1–60.

Page, D. N. and Geilker, C. (1981), "Indirect evidence for quantum gravity," *Physical Review Letters* **47**(14), 979.

Penrose, R. (1994), "Non-locality and objectivity in quantum state," in J. S. Anandan, J. L. Safko,. eds., *Quantum coherence and reality: In celebration of the 60th birthday of Yakir Aharonov-Proceedings of the International Conference On Fundamental Aspects of Quantum Theory. World Scientific*, pp. 238–246.

Pradeu, T., Lemoine, M., Khelfaoui, M. and Gingras, Y. (2021), "Philosophy in science: Can philosophers of science permeate through science and produce scientific knowledge?," *British Journal for the Philosophy of Science*, https://www.journals.uchicago.edu/doi/10.1086/715518.

Raju, S. (2022), "Failure of the split property in gravity and the information paradox," *Classical and Quantum Gravity* **39**(6), 064002.

Rosenfeld, L. (1963), "On quantization of fields," *Nuclear Physics* **40**, 353–356.

Roura, A. and Verdaguer, E. (2008), "Cosmological perturbations from stochastic gravity," *Physics Review D* **78**, 064010.

Rovenchak, A. and Krynytskyi, Y. (2018), "Radiation of the electromagnetic field beyond the dipole approximation," *American Journal of Physics* **86**(10), 727–732.

Rubino, G., Rozema, L. A., Feix, A. et al. (2017), "Experimental verification of an indefinite causal order," *Science Advances* **3**(3), e1602589.

Rydving, E., Aurell, E. and Pikovski, I. (2021), "Do Gedanken experiments compel quantization of gravity?," *Physical Review D* **104**(8), 086024.

Salecker, H. and Wigner, E. (1997), "Quantum limitations of the measurement of space-time distances," A. S. Wightman, ed., in *Part I: Particles and Fields. Part II: Foundations of Quantum Mechanics*, Springer, pp. 148–154.

Thébault, K. P. (2016), "What can we learn from analogue experiments?," *arXiv preprint arXiv:1610.05028*.

Vedral, V. (2006), "Witnessing quantum entanglement," in *Introduction to Quantum Information Science*, Oxford University Press.

Visser, M. (2002), "Sakharov's induced gravity: A modern perspective," *Modern Physics Letters A* **17**(15n17), 977–991.

Wallace, D. (2022), "Quantum gravity at low energies," *Studies in History and Philosophy of Science* **94**, 31–46.

Weatherall, J. O. (2014), "What is a singularity in geometrized Newtonian gravitation?," *Philosophy of Science* **81**(5), 1077–1089.

Weatherall, J. O. and Manchak, J. B. (2014), "The geometry of conventionality," *Philosophy of Science* **81**(2), 233–247.

Zimmermann, F. (2018), "Future colliders for particle physics—'big and small,'" *Nuclear Instruments and Methods in Physics Research Section A: Accelerators, Spectrometers, Detectors and Associated Equipment* **909**, 33–37.

Acknowledgments

Many people have provided us with invaluable feedback, error correction, and lengthy discussion, especially: Emily Adlam, Charis Anastopoulos, Markus Aspelmeyer, Marios Christodoulou, Richard DeJong, Andrea Di Bagio, Flaminia Giacomini, Bei-Lok Hu, Siddharth Krishnan, Philip Manheim, Chiara Marletto, Carlo Rovelli, Daniel Sudarsky, Karim Thébault, and David Wallace. We disagree (to varying degrees) with many of them, but they never failed to be enlightening. Thank you.

We are extremely grateful to attentive audiences at the Center for Philosophy of Science, the University of Pittsburgh, the University of Warsaw, the University of Oxford, the Cambridge-LMU Black Holes Workshop at DAMTP, the Universitat Autònoma de Barcelona, the 2022 Philosophy of Science Association Conference, the London School of Economics Sigma Club, and the University of Bristol.

This Element was made possible through the support of all three authors by a grant from the John Templeton Foundation. The opinions expressed in this publication are those of the author(s) and do not necessarily reflect the views of the John Templeton Foundation.

In addition, Nick Huggett was supported by a Senior Visiting fellowship from the Center for Philosophy of Science at the University of Pittsburgh, and a Benjamin Meaker Distinguished Visiting Professorship from the University of Bristol. Mike Schneider was supported by a Postdoctoral Fellowship from the Center for Philosophy of Science at the University of Pittsburgh.

Cambridge Elements ☰

Foundations of Contemporary Physics

Richard Dawid
Stockholm University

Richard Dawid is Professor in the Philosophy of Science at Stockholm University and specialises in the philosophy of contemporary physics, particularly that of non-empirical theory assessment. In 2013 he published *String Theory and the Scientific Method* and in 2019 he co-edited a second book titled *Why Trust a Theory* (both published by Cambridge University Press).

James Wells
University of Michigan, Ann Arbor

James Wells is Professor in Physics at the University of Michigan, Ann Arbor, and his research specialises in high-energy theoretical physics, with a particular focus on foundational questions in fundamental physics such as gauge symmetries, CP violation, naturalness and cosmological history. He is a Fellow of the American Physical Society.

About the Series

Foundations in Contemporary Physics explores some of the most significant questions and discussions currently taking place in modern physics. The series is accessible to physicists and philosophers and historians of science, and has a strong focus on cutting-edge topics of research such as quantum information, cosmology, and big data.

Cambridge Elements ≡

Foundations of Contemporary Physics

Printed in the United States
by Baker & Taylor Publisher Services

Printed in the United States
by Baker & Taylor Publisher Services